ENGINEER'S TOOLKIT™

Overview

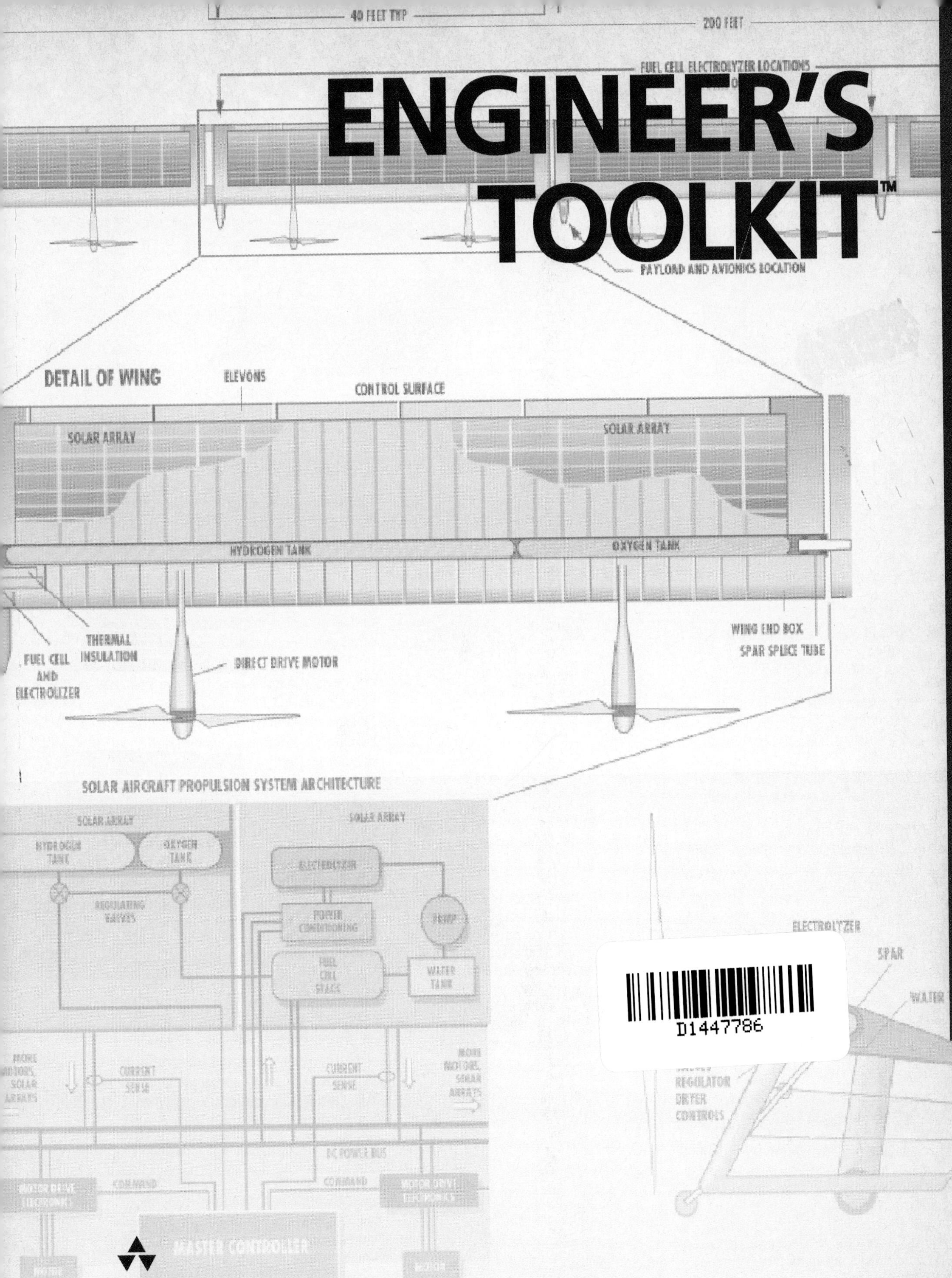

D1447786

Addison-Wesley Publishing Company, Inc.

Menlo Park, California · Reading, Massachusetts · New York · Don Mills, Ontario
Wokingham, U.K. · Amsterdam · Bonn · Paris · Milan · Madrid · Sydney
Singapore · Tokyo · Seoul · Taipei · Mexico City · San Juan, Puerto Rico

Executive Editor: Dan Joraanstad
Acquisitions Editor: Denise Penrose
Marketing Manager: Mary Tudor
Developmental Editors: Deborah Craig, Jeannine Drew, Kate Hoffman, Shelly Langman
Assistant Editor: Nate McFadden
Senior Production Editor: Teri Holden
Production Editors: Jean Lake, Gail Carrigan, Catherine Lewis
Supplements Production Editor: Teresa Thomas
Photo Editor: Lisa Lougee
Copy Editors: Barbara Conway, Robert Fiske
Proofreader: Holly McLean-Aldis
Marketing Coordinator: Anne Boyd
Cover Design: Yvo Riezebos
Text Design: Side by Side Studios
Technology Support: Craig Johnson
Composition: Side by Side Studios, Fog Press, London Road Design, Progressive, The Printed Page
Manufacturing Coordinator: Janet Weaver
Printing and Binding: R. R. Donnelley

Cover and Overview Photo Credits
Cover: Photo ©James Caccavo/Zuma Images; background illustration ©Ian Worpole
Photo 1: Courtesy of NASA
Photo 2: Courtesy of Jet Propulsion Laboratory
Photo 3: ©David Parker/SPL/Photo Researchers, Inc.
Photo 4: ©Brownie Harris/The Stock Market
Photo 5: Courtesy of M.E. Raichle, Wash. Univ., St. Louis
Photo 6: ©Chuck O'Rear/Westlight
Photo 7: Courtesy of Lockheed. Photo by Russ Underwood.
Photo 8: ©Roger Ressmeyer/Starlight
Photo 9: ©George Haling/Photo Researchers, Inc.
Photo 10: Courtesy of Keith Wood, Promega, Inc.

Copyright © 1996 by Addison-Wesley Publishing Company, Inc.

All rights reserved. No part of this publication may be reproduced, or stored in a database or retrieval system, distributed, or transmitted in any form or by any means, electronic, mechanical, photocopying, recording, or otherwise without the prior written permission of the publisher. Printed in the United States of America. Published simultaneously in Canada.

The programs, worksheets, and examples presented in this book have been included for their instructional value. They have been tested with care but are not guaranteed for any particular purpose. The publisher does not offer any warranties or representations, nor does it accept any liabilities with respect to the programs, worksheets, or examples.

ISBN: 0-8053-6335-1

Addison-Wesley Publishing Company, Inc.
2725 Sand Hill Road
Menlo Park, CA 94025

COVER STORY

Pictured on the cover of The Engineer's Toolkit is the Pathfinder—a "solar-powered flying wing" designed for low-speed, high-altitude flight. With a wing span comparable to a Boeing 737, it weighs in at just 400 pounds and has no rudders, no fins, no tail, and no pilot! The Pathfinder is one of a series of solar planes developed and built by Dr. Paul MacCready and his team of engineers at AeroVironment Inc., in Simi Valley, California. Engineers at the Lawrence Livermore National Laboratory in Livermore, California, designed, engineered, and continue to administer the Pathfinder solar plane. This laboratory also is designing the next iteration of solar planes, the Helios (plans for which appear behind the photograph of the Pathfinder). With the Helios, engineers hope to come even closer to realizing the dream of "eternal flight"; it will include an on-board energy storage system that can provide the energy needed during night flight.

As with most contemporary engineering projects, designing solar planes requires the efforts of engineers from a variety of disciplines—aeronautical, computer, electrical, environmental, and mechanical, to name a few. Still other teams of engineers are needed to design on-board equipment to support specific missions, such as monitoring dangerous weather systems or tracking the release of toxins into the atmosphere.

Contemporary design examples such as these are presented throughout The Engineer's Toolkit, highlighting the interdisciplinary teamwork that characterizes engineering today.

Tools for a New Curriculum

The Engineer's Toolkit is not a conventional textbook. It was inspired by the needs of instructors like you, who are engaged in developing a new curriculum in introductory engineering courses. They are searching for new ways to prepare, motivate, and engage first-year students. They want to create a link for their students between the prerequisite math and science courses and the wide range of engineering disciplines that build on that knowledge. These instructors also want to ensure that their students master the skills of team-building, communications, and computer use—skills they need to solve problems successfully in subsequent courses and in the real world of work. You and your colleagues are also experimenting with hands-on design projects so students understand that design is a process and that, fundamentally, engineering means solving problems.

Universities and colleges are responding in unique ways to the changing landscape of introductory engineering. This very uniqueness creates a new challenge when you are searching for the right text to support your unique course. *The Engineer's Toolkit* takes on that challenge. You choose from a rich set of course materials that introduce fundamental concepts of engineering and teach essential skills and tools. Each tool is presented as a single module. You determine which modules will best satisfy your course goals, and Addison-Wesley binds those modules into the exact book your students need.

Especially written and designed for *The Engineer's Toolkit*, the modules present a consistent teaching methodology adapted from the work of Delores Etter, author of the spreadsheet and Fortran *Toolkit* modules, as well as *Structured Fortran 77 for Scientists and Engineers.* Each of the *Toolkit* authors has applied Dr. Etter's five-step problem-solving process to a wide variety of programming languages and application programs. A consistent approach, style, level, and tone means you and your students don't have to switch gears every time you begin to teach or learn a new tool.

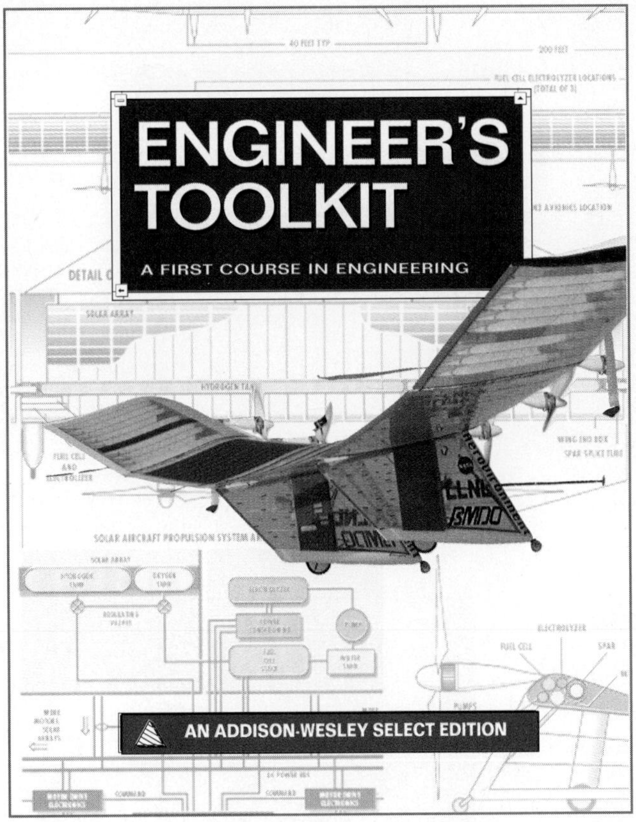

Here are the six key pedagogical features you'll find in the *Toolkit* modules that teach programming languages or software tools:

- **The five-step problem-solving process** is explained and illustrated in terms of the particular language or software tool being taught. It is then used throughout the module in applications, numbered examples, and end-of-chapter exercises or problems.
- **Applications** based on the Ten Great Engineering Achievements and representing a wide variety of engineering disciplines demonstrate the five-step problem-solving process.
- **"What If?" problems** immediately follow applications in the software tools modules, asking students to modify assumptions, data, or variables in the application and to solve the new problems that result.
- **Numbered examples** demonstrate key elements of a language or application program by providing fully worked-out solutions.
- **"Try It!" exercises** test students' knowledge of sections within a chapter and frequently require work at the computer.
- **End-of-chapter material** includes summaries of essential points, a key word list, and a set of exercises or problems that gradually increase in complexity.

These pedagogical features are also described from a student's point of view in the section "The *Toolkit* Methodology."

How To Design Your Custom Textbook

A sample course goal: To introduce engineering, teach a programming language, word processing and CAD techniques.

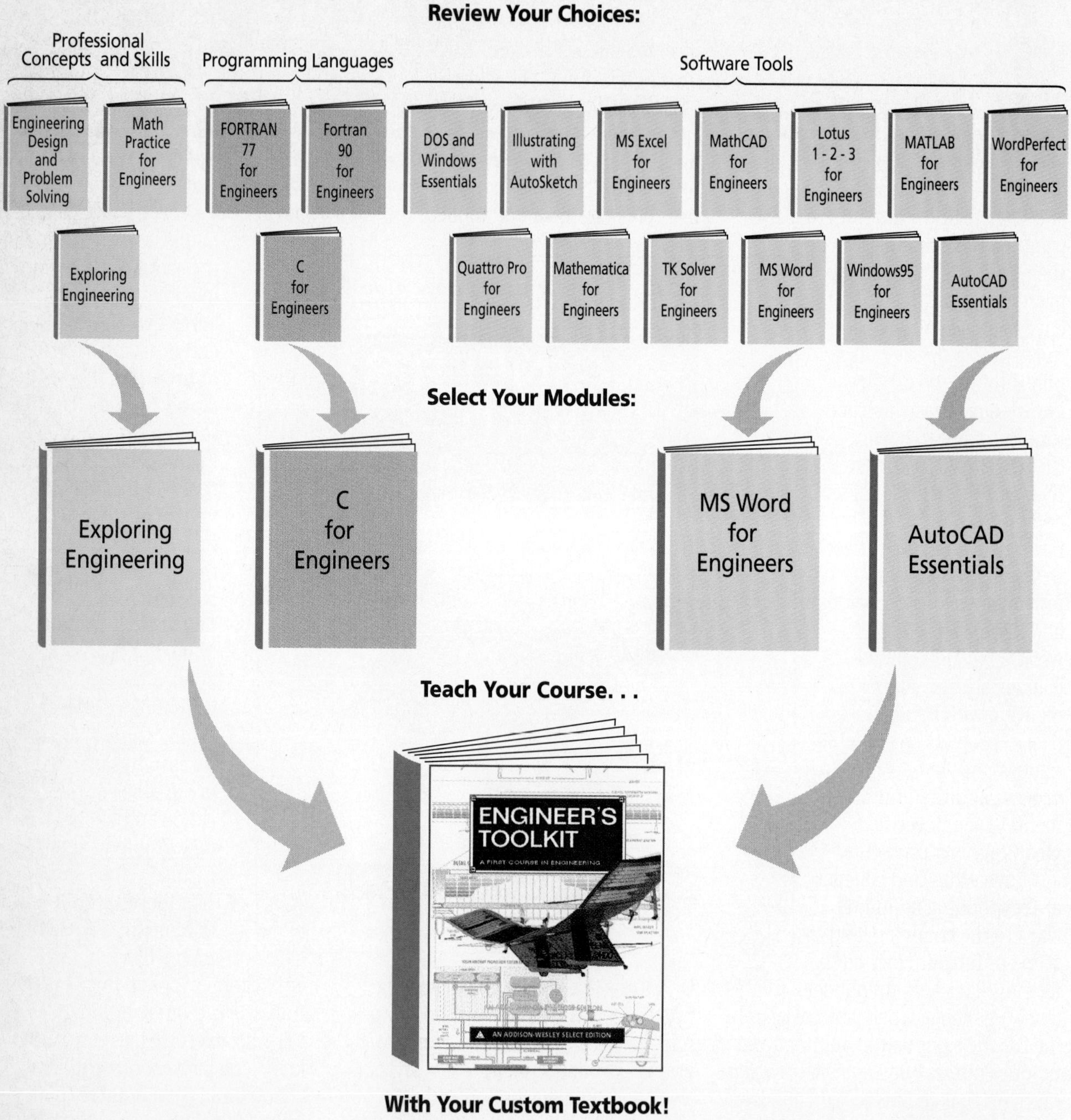

DESIGN YOUR OWN TOOLKIT

The *Toolkit* menu consists of three modules that teach programming languages, thirteen modules that teach software tools, and three modules that focus on core engineering concepts and skills—*Engineering Design and Problem Solving, Exploring Engineering,* and *Math Practice for Engineers.* You can mix and match as your course demands.

In a course that combines an overview of engineering disciplines with essential computer applications and the basic design process, you might, for example, combine *Exploring Engineering, Engineering Design and Problem Solving, Quattro Pro for Engineers,* and *Microsoft Word for Engineers* for a book of about 460 pages.

If you plan to teach programming and introduce your students to design and basic CAD techniques, you might combine *Engineering Design and Problem Solving, C for Engineers,* and *AutoCAD Essentials* for a book of about 550 pages.

The *Toolkit* modules are also available separately, and each custom text includes this Overview, which introduces instructors and students to the *Toolkit* methodology.

FUTURE MODULES

Each year Addison-Wesley will release additional modules to keep pace with the development in introductory engineering courses. We welcome your suggestions for future modules in *The Engineer's Toolkit.* Please correspond with us at the Internet address

toolkit@aw.com

or by regular mail at

Toolkit
Addison-Wesley Publishing Company, Inc.
2725 Sand Hill Road
Menlo Park, CA 94025

Toolkit information is also available on the World Wide Web at

http://www.aw.com/cseng/toolkit/

ELECTRONIC SUPPLEMENTS AND SUPPORT MATERIAL

Addison-Wesley offers a range of support materials for *The Engineer's Toolkit,* both printed and electronic. We welcome your comments on the effectiveness of these materials and your suggestions for additional supplements.

Supplements via the Internet A selection of supplementary material for *The Engineer's Toolkit,* including transparency masters, is available on the Internet via anonymous FTP. The URL for accessing this material is ftp://aw.com/cseng/toolkit.

Instructor's Guides Instructor's Guides for the modules you choose will be provided upon adoption of *The Engineer's Toolkit.*

Each Instructor's Guide opens with an overview of the module subject matter, suggested methods of instruction, comments on tests and quizzes, and a general discussion of software version and platform issues (if applicable).

Each Instructor's Guide also describes how to order the module topics to support various syllabi.

A chapter-by-chapter section presents teaching strategies, points to emphasize and special challenges for each chapter, solutions to the end-of-chapter problems, and additional problems and solutions.

Student Data Files Student Data Files, including the data, applications, and program files that support end-of-chapter problems in specific modules, will be available in the following ways:

- on the disk that includes the Instructor's Guide
- via our FTP site: ftp://aw.com/cseng/toolkit/igs/
- via our Toolkit home page on the World Wide Web: http://www.aw.com/cseng/toolkit/

Modules with Data Files include *AutoCAD Essentials, Illustrating with AutoSketch, FORTRAN 77 for Engineers, Fortran 90 for Engineers, C for Engineers, Quattro Pro for Engineers, Lotus 1-2-3 for Engineers, Microsoft Excel for Engineers, TK Solver for Engineers,* and *Mathematica for Engineers.*

Errata for Published Modules: Errata notices for published modules will be available online at: http://www.aw.com/cseng/toolkit/

FIRST ASSIGNMENT

The next section of this Overview is directed to your students. It explains the teaching and learning strategies adopted by our authors throughout the *Toolkit.* The final section introduces students to the Ten Great Engineering Achievements. We invite you to read on and hope you will assign this section to your students early in your course.

THE ENGINEER'S TOOLKIT MODULES

	Title	Author
Professional Concepts and Skills	Engineering Design and Problem Solving	Steve Howell
	Exploring Engineering	Joe King
	Math Practice for Engineers	Joe King
Software Tools	AutoCAD Essentials	Melton Miller
	DOS and Windows Essentials	Gerald Lemay
	Illustrating with AutoSketch	Gordon Snyder
	Lotus 1-2-3 for Engineers	Delores Etter
	MathCAD for Engineers	Joe King
	Mathematica for Engineers	Henry Shapiro
	MATLAB for Engineers	Joe King
	Microsoft Excel for Engineers	Delores Etter
	Microsoft Word for Engineers	Sheryl Sorby
	Quattro Pro for Engineers	Delores Etter
	TK Solver for Engineers	Robert J. Ferguson
	Windows95 for Engineers	Gordon Snyder
	WordPerfect for Engineers	Sheryl Sorby
Programming Languages	C for Engineers	Kenneth Collier
	FORTRAN 77 for Engineers	Delores Etter
	Fortran 90 for Engineers	Delores Etter

The Toolkit Methodology

Welcome to *The Engineer's Toolkit!* This book has been especially created to support your work in what is probably your first course in Engineering. Unlike other textbooks you have studied, *The Engineer's Toolkit* was customized by your instructor to include the exact material you need, and only the material you need. In essence, *The Engineer's Toolkit* is a collection of modules that teach engineering concepts and skills, software tools, and programming languages. Your instructor has selected the appropriate modules for your course, and Addison-Wesley has bound those modules into this custom book.

A GENERAL PROCESS FOR SOLVING PROBLEMS

A key feature of *The Engineer's Toolkit* is its emphasis on developing problem-solving skills. Problem solving is one of the foundations of all engineering activity. In *The Engineer's Toolkit* you'll find a five-step method for solving the problems given in each module. Some engineers will tell you they use a nine-step process; others can condense their process down to four. There's nothing magic about the number, but you will find that learning and following a consistent method for solving problems will make you an efficient student and a promising graduate. Each application program or programming language module builds on this general problem-solving method:

1. Define the problem.
2. Gather information.
3. Generate and evaluate potential solutions.
4. Refine and implement the solution.
5. Verify the solution through testing.

Introduction. Each application is fully described and explained so that you have sufficient information to complete step 1.

66 QUATTRO PRO FOR ENGINEERS

and choosing it. An icon will be on this page for each graph book. Choose the icon of the graph you want to rename with and select Properties|Current Object. The Name dialog box t in which you may type the new name of the graph and then c

Saving and printing graphs is similar to saving and prin sheets. The graph is automatically saved when you save the file. You can print the graph either by itself or with the spre print the graph by itself, select the graph using Graph|Ed choose File|Print. If you have placed the graph as a floating g spreadsheet, it is printed when you print the spreadsheet.

Try It Change the name of the GRAPH1 graph in the FILTER1 spreads NALIN. Then print the graph.

Application 1 QUALITY CONTROL

Manufacturing Engineering

In a manufacturing or assembly plant, quality control receives tion. One of the key responsibilities of a quality control engin lect accurate data on the quality of the product being manufa data can be used to identify the problem areas in the assemb the materials being used in the product.

Circuit Board Defects

In this application, information collected over a one-year peri specify both the type of defects and the number of defects de assembly of printed circuit boards. These defects have been four categories: board errors, chip errors, processing errors, tion errors. Board errors are typically caused by defects in th of the printed circuits. Chip errors are caused by defective in cuit (IC) chips that are added to the board; these IC chips incl chips, microprocessor chips, and digital filter chips. Processi typically caused by errors in inserting the chips in the board; is often done by manufacturing robots, and the robot progr be incorrect, or the chips being inserted can be packaged i order. Connection errors are solder errors that occur when th through the solder machine; these errors can be caused by board or an incorrect solder temperature.

Spreadsheet for a Quality Analysis Report

You want to develop a spreadsheet that summarizes the qu data that has been collected each month for a year. This data number of defects in each of the four categories of defects di summary report should compute totals and percentages fo year and defect totals for each quarter. Later in this chapter the data in the spreadsheet to generate pie, bar, and line grap

From *Quattro Pro® for Engineers*

184 C FOR ENGINEERS

```
void init_array(struct sample_type array[MAX_RAINFALLS][MAX_SITES])
{
  int row, col;        /* Loop control variables */

  for (row=0; row < MAX_RAINFALLS; ++row)
    for (col=0; col < MAX_SITES; ++col){
      array[row][col].date.day = 0;
      array[row][col].date.month = 0;
      array[row][col].date.year = 0;
      array[row][col].time.hour = 0;
      array[row][col].time.minutes = 0;
      array[row][col].h_concentration = 0.0;
      array[row][col].ph_level = 0.0;
    }
}
```

Carefully examine the definition of `init_array()`. Notice that the parameter array is declared as a two-dimensional array of structures. In the body of this function access is made to elements within `array` using the familiar subscript notation. Once a particular element has been accessed in this manner, the fields of that structure are accessed using the dot operator. Whenever the accessed field is itself a structure, its fields are accessed using a second dot operator, as in the statement

```
array[row][col].date.month = 0;
```

Because the dot is an operator, it can be combined with other operators in this way.

There are many powerful ways to use structures in a program, and this section provides you with only an introduction. In the following application you will learn how structures can be used to extend the mathematical power of *C* to include complex arithmetic.

Application 1 FREQUENCY DISTRIBUTION GRAPHING

Industrial Engineering

Quality-control engineers monitor the quality of an automated production line by tracking the number of defective parts coming off the line within a particular period. If the frequency of defective parts rises dramatically for a given period, the engineer is alerted that a problem exists and can take action to fix the problem. Such frequencies can be depicted using a bar graph such as the one shown in Figure 7-5. In this graph, the horizontal axis represents the number of defects detected and the vertical axis represents data collection periods.

From *C for Engineers*

CHAPTER 4 ENGINEERING GRAPHS 67

1. Problem Statement

Generate a summary of the quality control monthly data over the past year. Include totals and percentages of the four defect categories along with quarterly totals.

2. Input/Output Description

Begin the spreadsheet using a solution template containing a title, description, and the parameters of the spreadsheet. The monthly data represents the input parameters. Figure 4-11 contains a window of the report template.

Figure 4-11
Report Template

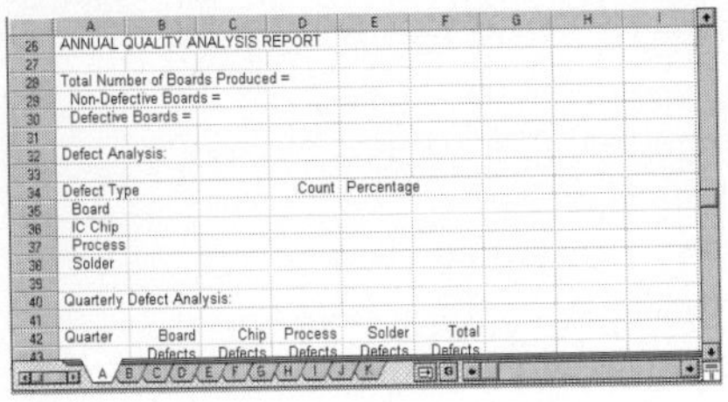

3. Hand Example

Generate a sample set of data:

	Boards Passed	Boards Rejected	Number of Defects			
			Board	Chip	Process	Solder
Jan	1201	34	8	12	4	10
Feb	890	15	3	5	2	5
Mar	933	24	13	6	0	5
Apr	1022	18	9	3	3	3
May	975	10	4	3	1	2
Jun	864	13	8	4	0	1
Jul	891	17	6	5	2	4
Aug	903	11	4	6	0	1
Sep	1075	18	10	3	4	1
Oct	1180	21	11	6	1	3
Nov	1380	34	20	11	0	3
Dec	903	11	4	2	3	2

From *Quattro Pro® for Engineers*

CHAPTER 7 COMPOSITE DATA TYPES 185

1. Define the Problem

The problem is how to write a program that will read the number of defective parts for each half-hour in a single day and store them as elements in a one-dimensional array. This array is to be used to display a frequency distribution graph on the computer's display monitor.

2. Gather Information

The input data for this program is a frequency count of defective parts for each half-hour beginning with 12:00 midnight on one day and running to 12:00 midnight of the following day, consecutively. Hence there will be 48 frequency counts, each of which will be a single integer value.

3. Generate and Evaluate Potential Solutions

The problem can be decomposed into two general parts: reading frequency data and displaying the frequency graph. Although there are 48 data collections in a given day, it is possible that frequency data is available for only some of these periods. So, reading frequency data requires the following steps:

```
Read and check number of periods for which data is available
Read and check each frequency data value
```

The structure chart shown in Figure 7-6 reflects the design of a program to solve this problem as shown below:

```
Do
   Prompt user for number of data values
   Read number
While number < 0 or > 48
```

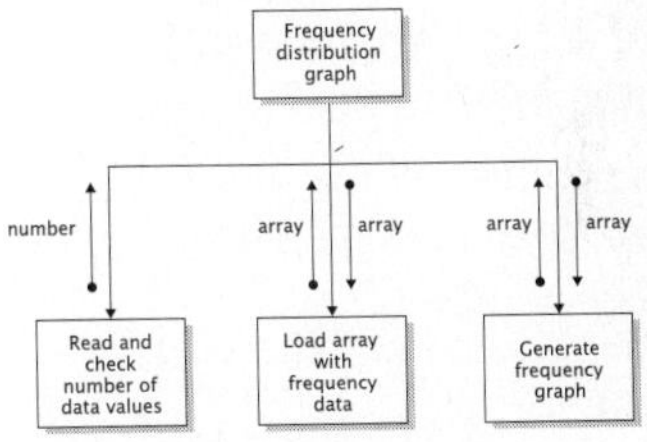

Figure 7-6 Structure Chart for Distribution Plotting

From *C for Engineers*

Step 1. In both applications you are asked to **define the problem.** The introductory description gives lots of clues to help you.

Step 2. This step asks you to **gather information** that you need to propose a solution. In these applications you need to prepare the data that will be used to generate quality analysis reports.

Step 3. You are now ready to **generate and evaluate potential solutions.** In C you create a structure chart to illustrate the design of a program that can solve the problem. In Quattro Pro, the hand example shows how the gathered data will be used to determine the algorithm in step 4.

A SPECIFIC FIVE-STEP PROCESS

Each module adapts this general method and refines it according to the kinds of problems solved by the tool or language being taught. Chapter 1 of each programming language and application program module describes the specific five-step process used in that module.

As you work through *The Engineer's Toolkit*, you'll find that this consistent approach makes it easier to solve new problems. For instance, step 1 of the five-step problem-solving process calls for the same kind of thinking process whether you are using a programming language like Fortran 90, a computer-aided design package like AutoCAD, or an equation solver like MATLAB.

We illustrate the five-step problem-solving process with a pair of applications from *The Engineer's Toolkit*. Both of the examples presented here deal with the collection and tracking of data related to quality control. Each has been fully worked out using the five-step process. Follow these steps to see how easy it is to learn this problem-solving process.

Step 4. In this step you write a C program based on the structure chart and algorithms developed in the previous steps. You will now **refine and implement the solution.** With Quattro Pro, this step means developing the formulas that will be used to compute the values listed in the summary report.

Step 5. In the final step you **verify the solution through testing.** In C this involves entering a variety of values (value testing) to confirm that the program generates valid output. In Quattro Pro the spreadsheet is tested with several sets of data to verify the accuracy of the computations. Accuracy is confirmed by comparing the spreadsheet calculations to the values determined in the hand example.

APPLICATIONS

These two applications are among hundreds in which *Toolkit* authors demonstrate the five-step problem-solving method. As you gain experience using this method with various software tools and languages, you'll find you can approach new problems with confidence, and you'll begin to identify the appropriate tool or language for the problem at hand. Learning to choose the right tool for a specific engineering problem is an important part of your education.

Many of the applications in *The Engineer's Toolkit* are based on the Ten Great Engineering Achievements chosen by the National Academy of Engineering to celebrate its silver anniversary in 1989. Studying these applications will help you see the kinds of problems faced by engineers from different disciplines and better understand how large problems are broken into smaller solvable problems. This Overview concludes with a description of the Ten Great Engineering Achievements.

68 QUATTRO PRO FOR ENGINEERS

4. Algorithm Development

The spreadsheet now contains everything except the formulas for computing the summary information for the report. Develop the formulas in the order needed to compute the values. Also, try to minimize the number of computations. For example, since you generate error sums by quarters, add the quarterly sums to get yearly sums instead of adding all the monthly sums to get yearly sums. It would be good practice to verify each of these formulas by referring to Figures 4-10 and 4-11.

D29	@SUM(D11..D22)	Total non-defective boards
D30	@SUM(E11..E22)	Total defective boards
E28	+D29+D30	Total boards
E29	+D29/E28	Percent non-defective boards
E30	Copy from E29	Percent defective boards
B44	@SUM(G11..G13)	Quarter I board defects
B45	@SUM(G14..G16)	Quarter II board defects

5. Testing

An important part of developing a spreadsheet is testing it with several sets of data to verify the accuracy of the computations. Using the sample set of data from the hand example, you can easily check the accuracy of the spreadsheet calculations by comparing them to the hand example.

You should make minor changes in this data and check the report to be sure that corresponding changes occur in the report summary. In this report you want to be sure that the report would be generated correctly if there were no errors in one of the categories. Be sure to change the corresponding sums of boards rejected. The corresponding report generated, shown in Figure 4-12, shows that there were no process defects during any of the four quarters.

Figure 4-12
Report with No Processing Defects

ANNUAL QUALITY ANALYSIS REPORT					
Total Number of Boards Produced =			12,423		
Non-Defective Boards =		12,217	98.34%		
Defective Boards =		206	1.66%		
Defect Analysis:					
Defect Type		Count	Percentage		
Board		100	48.54%		
IC Chip		66	32.04%		
Process		0	0.00%		
Solder		40	19.42%		
Quarterly Defect Analysis:					
Quarter	Board Defects	Chip Defects	Process Defects	Solder Defects	Total Defects
I	24	23	0	20	67
II	21	10	0	6	37
III	20	14	0	6	40
IV	35	19	0	8	62

From *Quattro Pro® for Engineers*

C FOR ENGINEERS

4. Refine and Implement a Solution

The structure chart and algorithms developed in the previous section are implemented as the program in Example 7 10.

5. Verify and Test the Solution

To properly test this program, you should enter a variety of values for first input, including 0, 48, values below 0, values above 48, and valid values. Selecting values along the boundaries is known as boundary value testing. You should do the same for the actual frequency values. Given the input values −3 (error) 55 (error) 10−9 (error) 15 16 13 14 20 9 96 (error) 19 14 19 20, the output of this program is

```
     -------------------------------------------------
 1 | ***************
 2 | ****************
 3 | *************
 4 | **************
 5 | ********************
 6 | *********
 7 | *******************
 8 | **************
 9 | *******************
10 | ********************
     --- 5---10---15---20---25---30---35---40---45---50
```

This chapter introduced you to one-, two- and three-dimensional arrays, as well as structures. Arrays and structures offer a cohesive means of storing composite collections of data. Arrays are used for grouping items of the same type and meaning, whereas structures allow you to group related items with different types and meanings. You learned how to declare and initialize arrays and manipulate the elements in an array. You also learned how to define a structure, declare structure variables, and manipulate the fields in a structure. Finally, you learned that these composite data types can be combined to allow you to create arrays of structures, structures containing arrays, and structures containing other structures.

From *C for Engineers*

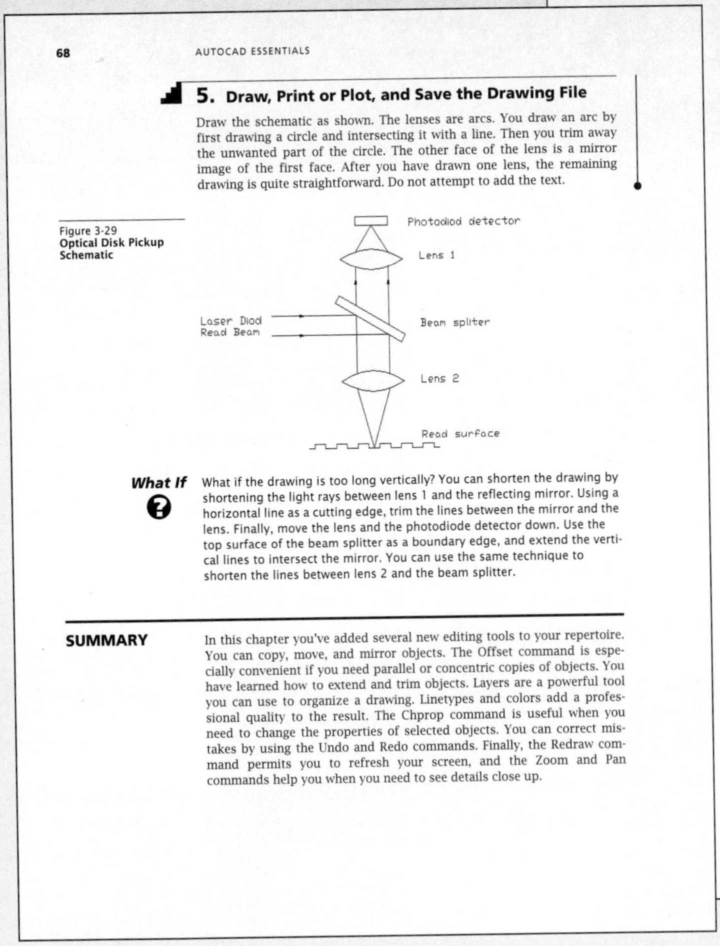
68 AUTOCAD ESSENTIALS

5. Draw, Print or Plot, and Save the Drawing File

Draw the schematic as shown. The lenses are arcs. You draw an arc by first drawing a circle and intersecting it with a line. Then you trim away the unwanted part of the circle. The other face of the lens is a mirror image of the first face. After you have drawn one lens, the remaining drawing is quite straightforward. Do not attempt to add the text.

Figure 3-29
Optical Disk Pickup Schematic

What If What if the drawing is too long vertically? You can shorten the drawing by shortening the light rays between lens 1 and the reflecting mirror. Using a horizontal line as a cutting edge, trim the lines between the mirror and the lens. Finally, move the lens and the photodiode detector down. Use the top surface of the beam splitter as a boundary edge, and extend the vertical lines to intersect the mirror. You can use the same technique to shorten the lines between lens 2 and the beam splitter.

SUMMARY In this chapter you've added several new editing tools to your repertoire. You can copy, move, and mirror objects. The Offset command is especially convenient if you need parallel or concentric copies of objects. You have learned how to extend and trim objects. Layers are a powerful tool you can use to organize a drawing. Linetypes and colors add a professional quality to the result. The Chprop command is useful when you need to change the properties of selected objects. You can correct mistakes by using the Undo and Redo commands. Finally, the Redraw command permits you to refresh your screen, and the Zoom and Pan commands help you when you need to see details close up.

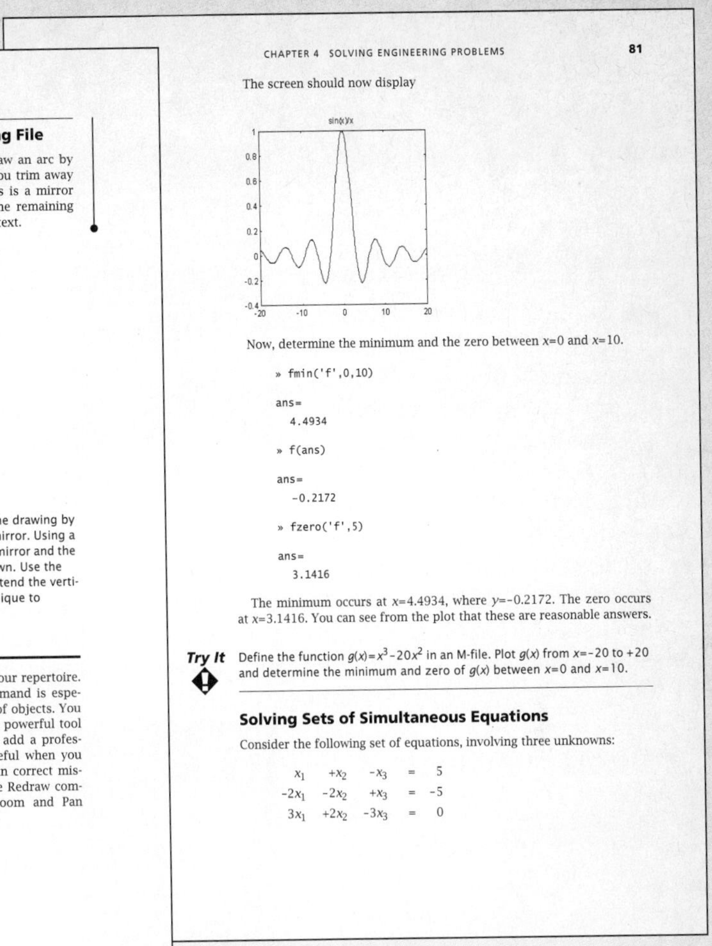
CHAPTER 4 SOLVING ENGINEERING PROBLEMS 81

The screen should now display

Now, determine the minimum and the zero between x=0 and x=10.

```
» fmin('f',0,10)

ans=
   4.4934

» f(ans)

ans=
   -0.2172

» fzero('f',5)

ans=
   3.1416
```

The minimum occurs at x=4.4934, where y=−0.2172. The zero occurs at x=3.1416. You can see from the plot that these are reasonable answers.

Try It Define the function $g(x)=x^3-20x^2$ in an M-file. Plot $g(x)$ from x=−20 to +20 and determine the minimum and zero of $g(x)$ between x=0 and x=10.

Solving Sets of Simultaneous Equations

Consider the following set of equations, involving three unknowns:

$$\begin{aligned} x_1 + x_2 - x_3 &= 5 \\ -2x_1 - 2x_2 + x_3 &= -5 \\ 3x_1 + 2x_2 - 3x_3 &= 0 \end{aligned}$$

MATLAB® for Engineers

AutoCAD® Essentials

ENGINEERING DISCIPLINES

In *AutoCAD Essentials*, the author presents applications from mechanical and materials engineering, civil and structural engineering, electrical and electronics engineering, and optical engineering. *MATLAB for Engineers* includes applications from electrical and computer engineering. You can find the listing of application problems in Chapter 1 of each software tool or programming language module. The table below lists several applications from the FORTRAN 77 module.

Applications	**ACROSS THE DISCIPLINES**	**Chapter**
Stride Estimation	Mechanical Engineering	2
Light Pipes	Optical Engineering	3
Sonar Signals	Acoustical Engineering	4
Wind Tunnels	Aerospace Engineering	5
Oil Well Production	Petroleum Engineering	6
Simulation Data	Electrical Engineering	7

"What If" Problems

These problems immediately follow an application. Often you are asked to test the model developed in the application with different data or different assumptions. Working through these problems ensures you fully understand the application.

"Try It!" Exercises

Each chapter contains several "Try It!" exercises. A "Try It!" is a short set of exercises that tests your understanding of the material. These exercises increase in complexity over the course of each module, and if you try out each one, you'll find you'll master the material faster than you expected.

NUMBERED EXAMPLES

Numbered examples appear in all the programming language and some software tool modules. These examples are designed to illustrate specific features of the language or software. Working through these examples is essential, especially those that offer two solutions and a discussion of the differences between them.

EXERCISES/PROBLEMS

Every module includes end-of-chapter exercises or problems that increase in difficulty and test your knowledge of the chapter material. Make sure you try your hand at the excercises that require you to use the five-step method to find solutions.

The *Toolkit* Team

As you consider the Ten Great Engineering Achievements explored at the end of this Overview, you'll notice that many require contributions from several different engineering disciplines. Today, significant projects can only be accomplished by teams of professionals. And that's true of *the Engineer's Toolkit,* too. The team that has helped to create the *Toolkit* includes not only the authors but also the focus group participants who honed and directed the *Toolkit* concept and the reviewers who helped develop the individual manuscripts.

TOOLKIT AUTHORS

Delores Etter, author of five modules, *FORTRAN 77 for Engineers, Fortran 90 for Engineers, Lotus 1-2-3 for Engineers, QuattroPro for Engineers, and Microsoft Excel for Engineers,* is a professor of electrical and computer engineering at the University of Colorado at Boulder. Dr. Etter has helped shape *The Engineer's Toolkit* from its initial conception, contributing the five-step problem-solving process and key pedagogical features that were successfully tested in her earlier Addison-Wesley texts, such as *Structured FORTRAN 77 for Engineers and Scientists, Lotus 123: A Software Tool for Engineers,* and *Quattro Pro: A Software Tool for Engineers.*

Ken Collier, Assistant Professor of Computer Science in the Department of Computer Science and Engineering at Northern Arizona University in Flagstaff, is the author of *C for Engineers.* Professor Collier teaches courses in C, C++, software engineering, and engineering design. His areas of research include software engineering, software design methodologies, and artificial intelligence.

R. J. Ferguson, the author of *TK Solver for Engineers,* is a professor of mechanical engineering at the Royal Military College of Canada. He teaches courses in stress analysis and computer-aided design. Other publications by Professor Ferguson include texts in the fields of fracture mechanics, noncircular gearing, vehicle transmissions, and engineering education.

Steve Howell, Associate Professor of Engineering in the mechanical engineering department at Northern Arizona University in Flagstaff, Arizona, is the author of *Engineering Design and Problem Solving.* He teaches a course titled Introduction to Engineering Design and Graphics. Professor Howell's areas of research are computer-aided design and manufacturing, thermodynamics, and heat transfer.

Joe King, the author of *MathCAD for Engineers, MATLAB for Engineers, Exploring Engineering,* and *Math Practice for Engineers,* is an associate professor of electrical engineering at the University of the Pacific in Stockton, California. He teaches courses in electrical engineering, advanced digital design, local area networks, neural networks, machine vision, C++, and microprocessor applications. He conducts research in the areas of neural networks and microprocessor applications.

Gerald Lemay, Professor of Electrical and Computer Engineering at the University of Massachusetts, Dartmouth, is the author of *DOS and Windows Essentials.* He teaches the Science of Engineering for honors students and Computer Tools for Engineers. Professor Lemay does research in renewable energy.

Melton Miller, author of *AutoCAD Essentials,* is an associate professor of civil engineering and assistant dean of the College of Engineering at the University of Massachusetts, Amherst. He teaches courses in Pascal, Fortran, Lotus, MathCAD, and AutoCAD. He also teaches courses in the the design of reinforced concrete structures.

Henry Shapiro is the author of *Mathematica for Engineers.* He is an associate professor of computer science at the University of New Mexico, where he teaches courses in computer programming and mathematical foundations of computer science. Professor Shapiro conducts research in the area of algorithm design. He is also active in curriculum development and accreditation of computer science programs.

Sheryl Sorby is the author of *WordPerfect for Engineers* and *Microsoft Word for Engineers.* She is an assistant professor of civil and environmental engineering at Michigan Technological University in Houghton, where she teaches courses in freshman engineering and computer skills. Professor Sorby conducts research in structural engineering.

Gordon Snyder, author of *Illustrating with AutoSketch* and *Windows 95 Essentials,* is an associate professor and department co-chair of the departments of electronics systems engineering, computer systems engineering, and laser electro-optics technology at Springfield Technical Community College in Springfield, Massachusetts. He teaches a course titled Introduction to Computer-Aided Engineering Technology.

REVIEWERS AND FOCUS GROUP PARTICIPANTS

Instructors throughout the country attended focus groups to help us identify key trends in engineering education. Over 100 reviewers contributed to the development of manuscripts for *The Engineer's Toolkit.* We gratefully acknowledge all their contributions.

Teresa Adams, University of Wisconsin-Madison · Anjum Ali, Mercer University · Abbas Aminmansour, Pennsylvania State University · A. Adnan Aswad, University of Michigan-Dearborn · Stormy Attaway, Boston University · James E. Bailey, Arizona State University · Betty Barr, University of Houston · O. Barron, University of Tennessee at Martin · Charles Beach, Florida Institute of Technology · Bill Beckwith, Clemson University · Charlotte Behm, Mission College · Lynn Bellamy, Arizona State University · Bhushan L. Bhatt, Oakland University · Steve Borgelt, University of Missouri · Joseph M. Bradley, United States Navy · Robert Brannock, Macon College · Gus Brar, Delaware Community College · David Bricker, Oakland University · Matthew Calame, Cosumnes River College · Christopher Carroll, University of Minnesota-Duluth · R. J. Coleman, University of North Carolina-Charlotte · James Collier, Virginia Polytechnic Institute and State University · Kenneth Collier, Northern Arizona University · Tom Cook, Mercer University · John Barrett Crittenden, Virginia Polytechnic Institute and State University · Barry Crittendon, Virginia Polytechnic Institute and State University · James Cunningham, Embry-Riddle Aeronautical University · Janek Dave, University of Cincinnati · Tim David, Leeds University · Al Day, Iowa State University · Jack Deacon, University of Kentucky · Bruce Dewey, University of Wyoming · Julie Ellis, University of Southern Maine · Beth Eschenbach, Humboldt State University · David Fletcher, University of the Pacific · Wallace Fowler, University of Texas-Austin · Jim Freeman, San Jose State University · Ben Friedman, Wolfram Research, Inc. · Paul Funk, University of Evansville · David Gallagher, Catholic University of America · Byron Garry, South Dakota State University · Larry Genalo, Iowa State University · Alan Genz, Washington State University · Johannes Gessler, Colorado State University · John J. Gilheany, Catholic University of America · Oscar R. Gonzalez, Old Dominion University · Yolanda Guran, Oregon Institute of Technology · Nicolas Haddad, New Mexico Institute of Mining and Technology · John Hakola, Hofstra University · Fred Hart, Worcester Polytechnic University · David Hata, Intel Corporation · Frank Hatfield, Michigan State University · Jim D. Helmert, Eastern Oklahoma State University · Herman W. Hill, Ohio University · Deidre A. Hirschfeld, Virginia Polytechnic Institute and State University · Margaret Hoft, University of Michigan-Dearborn · Henry Horwitz, Dutchess Community College · Peter K. Imbrie, Texas A & M University · Jeanine Ingber, University of New Mexico · Scott Iverson, University of Washington · Denise Jackson, University of Tennessee · Gerald S. Jakubowski, Loyola Marymount University · Scott James, GMI Engineering and Management Institute · Sundaresan Jayaraman, Georgia Institute of Technology · Johndan Johnson-Eilola, Purdue University · Douglas Jones, George Washington University · John Kelly, Arizona State University · Robert Knighton, University of the Pacific · Bill Koffke, Villanova University · Celal N. Kostem, Lehigh University · George V. Krestas, De Anza College · Tom Kurfess, Carnegie Mellon University · Ronald E. Lacey, Texas A & M University · Frank Lee, Bellevue Community College · Gerald J. Lemay, University of Massachusetts-Dartmouth · Jonathan Leonard, Saginaw Valley State University · Gloria Lewis, Wayne State University · John Lilley, University of New Mexico · L. N. Long, Pennsylvania State University · Robert A. Lucas, Lehigh University · Sharon Luck, Pennsylvania State University · Daniel D. Ludwig, Virginia Polytechnic Institute and State University · Arthur B. Maccabe, University of New Mexico · Jack Mahaney, Mercer University · Bill Marcy, Texas Tech University · Diane Martin, George Washington University · Bruce Maylath, University of Memphis · John McDonald, Rensselaer Polytechnic Institute · Olugbenga Mejabi, Wayne State University · Steve Melsheimer, Clemson University · Craig Miller, Purdue University · Howard Miller, Virginia Western Community College · Andrew J. Milne, The Leonhard Center for Innovation and Enhancement of Engineering Education · Pradeep Misra, Wright State University · Charles Morgan, Virginia Military Institute · F. A. Mosillo, University of Illinois · Stanley Napper, Louisiana Technical University · John Nazemetz, Oklahoma State University · James Nelson, Louisiana Technical University · Brad Nickerson, University of New Brunswick · Col. Kip Nygren, West Point Military Academy · Adebisi Oladipupo, Hampton University · Joseph Olivieri, Lawrence Technological University · Joseph Olson, University of South Alabama · Kevin Parfitt, Pennsylvania State University · John E. Parsons, North Carolina State University · Steve Peterson, Lawrence Livermore National Lab · Larry D. Piper, Texas A & M University · Jean Landa Pytel, Pennsylvania State University · Mulchand S. Rathod, Wayne State University · Philip Regalbuto, Trident Technical College · Thomas Regan, University of Maryland-College Park · Larry G. Richards, University of Maryland, College Park · Larry Riddle, Agnes Scott College · Lee Rosenthal, Fairleigh Dickinson University · Isidro Rubi, University of Colorado at Boulder · Vernon Sater, Arizona State University · Dhushy Sathianathan, Pennsylvania State University · Kenneth N. Sawyers, Lehigh University · Carolyn J. C. Schauble, University of Colorado at Boulder · William Schiesser, Lehigh University · Bruce Schimming, Howard University · Keith Schleiffer, Battelle · Murari J. Shah, Technical Graphics · Howard Silver, Fairleigh Dickinson University · Ed Simms, University of Massachusetts · Thomas Skinner, Boston University · Gary Sobczak, Purdue University · Jim Strumpff, Mercer University · Habib Taouk, University of Pittsburgh-Titusville · Massoud Tavakoli, GMI Engineering and Management Institute · Greg Taylor, Northern Arizona University · Ronald Terry, Brigham Young University · Ron Thurgood, Utah State University · Stephen Titcomb, University of Vermont · Joseph Tront, Virginia Polytechnic Institute and State University · Israel Ureli, Ohio University · Bert Van Grondelle, Hudson Valley Community College · Lambert VanPoolen, Calvin College · Akula Venkatram, University of California at Riverside · Thomas Walker, Virginia Polytechnic Institute and State University · Richard Wilkins, University of Delaware · Frazier Williams, University of Nebraska · Billy Wood, University of Texas-Austin · Nigel Wright, University of Nottingham

10 Great Engineering Achievements

In celebration of its silver anniversary, the U.S. National Academy of Engineering identified the Ten Great Engineering Achievements accomplished during the organization's first 25 years. Initially selected because they represent major breakthroughs, these achievements have initiated whole new areas of engineering. In the pages that follow, we note the types of design problems faced by interdisciplinary teams of engineers who work in these fields, and we represent contemporary examples in the photographs. Many of the applications you will encounter throughout *The Engineer's Toolkit* are based on these achievements.

1

SPACE TRAVEL

Although 1972 marked the last in a series of moon landings begun in 1969, the Apollo mission laid a foundation for a whole new generation of space shuttle missions that were dedicated to gathering information about the universe and that continue to test the human ability to travel in space.

Design Problems Several key design problems had to be solved to support the Apollo mission to land humans on the moon. The spacecraft required a new inertial navigation system; the lunar lander ascent engine had to work perfectly because there was no backup engine; the spacesuits had to protect the astronauts in a hostile environment and yet be flexible enough to allow movement; and the Saturn V rocket, which powered all the Apollo flights, had to be 15 times more powerful than the biggest rockets available in the early 1960s.

Application Areas Today, rockets launch deep space probes such as the Voyager, which continues to send back images from as far away as Venus and beyond, while space shuttles launch information-gathering satellites and scientific equipment. The Hubble Space Telescope (HST) was launched from the space shuttle Discovery on April 25, 1990. NASA had designed HST to allow scientists to view the universe up to 10 times more sharply than they could with earthbound telescopes. Unfortunately, scientists soon discovered that the primary mirror in HST was flawed and could not focus properly. It was another space shuttle, the Endeavour, that operated as a kind of Hubble repair station.

Engineering Disciplines Although the fully operational Hubble Space Telescope now stands as an impressive achievement for NASA, the repair job itself was perhaps an equally important achievement. Materials engineers helped develop metals, plastics, and other materials that could withstand the rigors of the launch and the environment in space. Mechanical engineers helped develop the mechanical structures that position HSTs mirrors and other moving apparatus and electrical engineers helped develop the complex computer, communication, and power systems.

The space shuttle Endeavour floats above the earth at an altitude of 381 miles, with the west coast of Australia forming the backdrop for the 35mm frame. While perched on top of a foot restraint on the Endeavour's Remote Manipulator System arm, astronauts F. Story Musgrove (top) and Jeffrey A Hoffman wrap up the last of five space walks. They have succeeded in their mission to repair the Hubble Space Telescope.

2

APPLICATION SATELLITES

Application satellites and other spacecraft orbit the earth to capture, relay, and transmit specific types of information, or to perform manufacturing processes that rely on special properties of the extraterrestrial environment, such as zero gravity.

Application Areas Satellite systems provide information on weather systems, relay communication signals around the globe, survey the earth and outer space to map uncharted terrain and provide navigational information for vehicles on land, in the oceans, and in the air. The Endeavour has been involved in NASA's Mission to Planet Earth, which is designed to help the international scientific community better understand which environmental changes are caused by nature and which are induced by human activity. Throughout 1994 the shuttle orbited the earth with the Spaceborne Imaging Radar-C and X-Band Synthetic Aperture Radar system, which illuminates the earth with microwaves, allowing detailed observations at any time, regardless of weather or sunlight conditions. (See photo below.)

Design Problems A satellite or spacecraft launching system must be designed to generate enough thrust to escape the earth's atmosphere. Once free, it needs to maintain a stable orbit around the earth. In addition, the hull needs to be light and yet strong enough to withstand the stress of the liftoff.

Engineering Disciplines Space-based inventions, such as the Spaceborne Imaging Radar-C (SIR), and satellites in general are a result of the cooperative efforts of aerospace engineers who help develop the systems that put satellites in space, and of chemical, mechanical, and electrical engineers who assist in the development of the radar and imaging systems for applications such as SIR.

This image of the area around Mount Pinatubo in the Philippines was acquired by the Spaceborne Imaging Radar-C and X-Band Synthetic Aperture Radar system aboard the space shuttle Endeavour in April 1994. This false color image shows the main volcanic crater on Mount Pinatubo produced by the June 1991 eruptions and the steep slopes on the upper flanks of the volcano. The red color shows the rougher ash deposited during the eruption. The dark drainages are the smooth mudflows that continue to flood the river valleys after heavy rains. This radar image helps identify the areas flooded by mudflows, which are difficult to distinguish visually, and assess the rate at which the erosion and deposition continue.

3

MICROPROCESSORS

A microprocessor is a tiny computer, smaller than your fingernail, that combines the control, arithmetic, and logic functions of large digital computers.

Application Areas With its small size and powerful capabilities, the applications of microprocessors range from operating remote television controllers or VCR recorders to providing the computational power in hand-held calculators or personal computers. Microprocessors can also be found in communication devices, such as networks that connect computers around the globe, and in automobiles, ships, and airplanes.

A technician working on the Bit Serial Optical Computer (BSOC), an optical computer that stores and manipulates data and instructions as pulses of light. To enable this, the designers (Harry Jordan and Vincent Heuring at the University of Colorado) developed bit-serial architecture. Each binary digit is represented by a pulse of infrared laser light 4 meters long. The pulses circulate sequentially through a tightly wound 4-kilometer loop of optical fiber some 50,000 times per second. Other laser beams operate lithium niobate optical switches which perform data processing.

Design Problems Key design problems involved in creating microprocessors include miniaturization, increasing speed while controlling the heat produced, and searching for materials stable and reliable enough to store, process, and transmit data at high speeds. For decades engineers have improved the performance of computers by increasing the number of functions contained on the CPU chip. Ultimately this approach created a bottleneck in switching between these functions. In the early 1980s, chip designers addressed this issue and developed a concept of Reduced Instruction Set Computing (RISC) which improves efficiency through the high-speed composition of an optimized minimal set of instructions. Research and development is on-going in another area of microprocessor design as well: optical computing. A dream since the 1940s, optical computing represents a fundamental change in how switching occurs—through optical signals rather than electronic signals. Since a computer is nothing more than a complex system of on/off switches, the speed at which the switches can turn off and on is the single most critical factor in determining the computer's performance. Engineers can design optical switches that operate well into gigahertz range, while electronic switches are currently restricted to about 100 megahertz.

Engineering Disciplines An application such as optical computing is a result of many years of collaboration between electrical engineers who design the complex laser systems, computer engineers who work on the computational structures, and chemical engineers who develop the actual lasers.

Inspection of the largest, most powerful energy-efficient jet engine, which was designed for the Boeing 777 jets.

4

JUMBO JET

Much of the success of jumbo jets (747, DC-10, L-1011) can be attributed to high-bypass fanjet engines, which allow the planes to fly farther using less fuel and with less noise than previous jet engines. Jumbo jets also have an increased emphasis on safety. For example, a 747 has four main landing-gear legs instead of two; a middle spar was added to the wings in the event one is damaged; and redundant hydraulic systems operate the critical system of elevators, stabilizers, and flaps that control the motion of the plane.

The newest jumbo jet is Boeing's 777, which was entirely developed using computer-aided design systems.

Application Areas The technological advances of superior fuels, engines, and materials achieved during the creation of the jumbo jet benefited the space program and the military.

Design Problems A major design problem faced by creators of the jumbo jet was creating an engine and fuel supply that could generate sufficient horsepower and thrust needed to lift the huge aircraft off the ground. The hull had to be strong enough to withstand the stress of the flight, without being so heavy that fuel efficiency was lost. The internal environment had to be comfortable for the passengers, especially during liftoff and landing.

Engineering Disciplines Engineers have been developing increasingly powerful jet engines ever since Englishman Sir Frank Whittle developed the first jet engine prototypes in 1937. The evolution of Whittle's engine into the Boeing 777 engine was largely due to the efforts of mechanical engineers who specialized in dynamics, thermodynamics, combustion systems, and materials.

5

MEDICAL IMAGING

A CAT (computerized axial tomography) scanner is a machine that generates three-dimensional images or slices of an object using x-rays. A series of x-rays is generated from many angles, encircling the object or patient. Each x-ray measures a density at its angle, and by combining these density measurements using sophisticated computer algorithms, an image can be reconstructed that gives clear, detailed pictures of the inside of the object. A PET (positron-emission tomography) scanner reveals locations of intense chemical activity within the body. Sugar labeled with radioactive isotopes that emit particles called positrons is first injected into a person's bloodstream. These positrons collide with electrons made available by chemical reactions in the body. The scanner detects the energy released by these collisions and maps metabolic "hot spots"—regions of an organ that are most chemically active at the time.

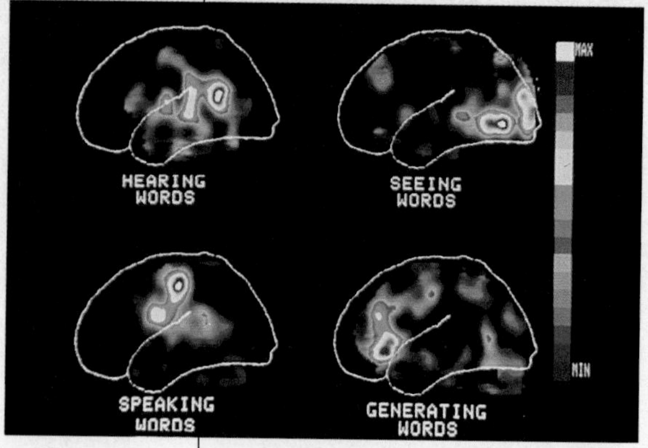

The PET scans shown here reveal localized brain activity under four different conditions, all related to language. Physicians use the PET scanner to diagnose brain and heart disorders and certain types of cancer.

Application Areas Doctors use diagnostic radiology to both detect and diagnose diseases, and in combination with other medical techniques, to treat diseases. These devices can also be used to explore the structure of both organic and inorganic materials.

Design Problems The key design problem lies in getting a clear picture without harming the patient or technician. In addition, the machine needs to withstand the various fields it produces.

Engineering Disciplines The PET and similar medical scanning systems are a product of biomedical engineering, a very specialized field of engineering that combines electrical engineering with medicine. Also involved are materials engineers—mechanical engineers who develop the special structures that contain, direct, and withstand the sometimes hazardous electromagnetic and radioactive emissions often used in such systems.

6

ADVANCED COMPOSITE MATERIALS

A composite consists of a matrix of one material that has been reinforced by the fibers or particles of another material. The choice of the composite is determined by the need of the application. For example, does the application require a material that is strong, flexible, stiff, lightweight, heavy, heat-tolerant, porous, dense, or wear-resistant?

Application Areas Advanced composites can be found in most products, including automobile components, communication systems, building materials, artificial joints and organs, and machine parts.

Design Problems Composite material designers must determine how the various materials will interact with each other and how the composites will act over time given the often high-stress situations in which they are used.

Engineering Disciplines Mechanical engineers are most frequently involved in the development of composite materials. Working with chemists and sometimes chemical engineers, mechanical engineers design applications, such as artificial limbs, for particular composite materials.

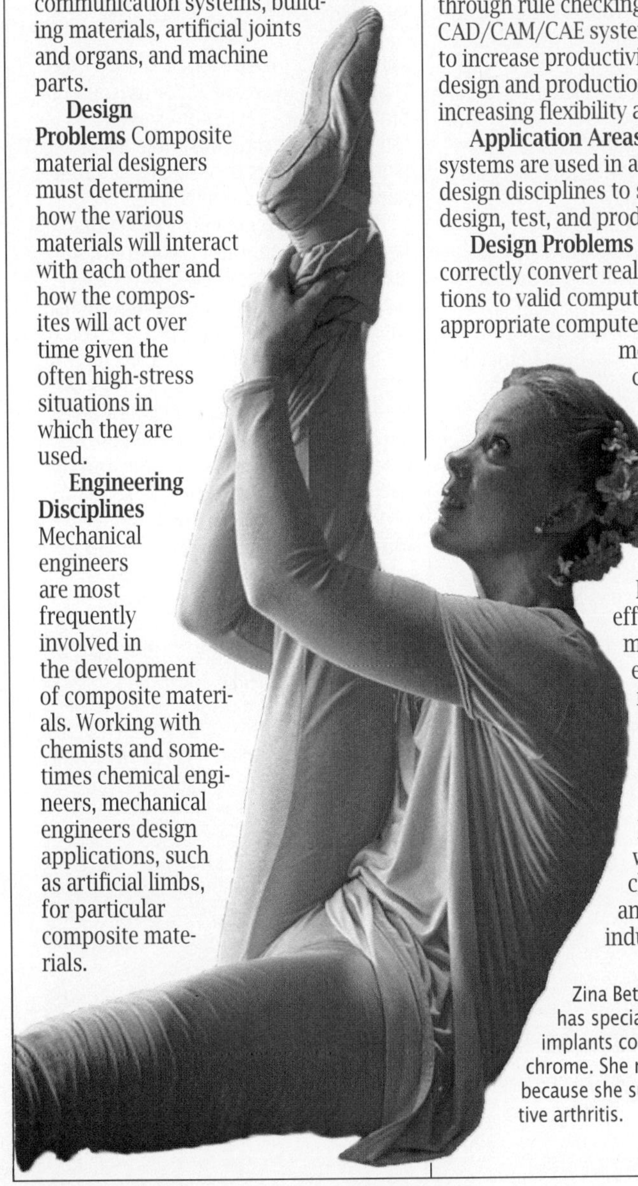

Zina Bethune, ballet teacher, has special long artificial hip implants coated with cobalt and chrome. She needed the implants because she suffers from degenerative arthritis.

7

CAD/CAM/CAE

CAD (computer-aided design) refers to computer systems used by engineers to model their designs. These systems may include plotters, computer graphics displays, and 2-D and 3-D modelers. CAM (computer-aided manufacturing) systems are used to control the machinery or industrial robots used in manufacturing the parts, assembling components, and moving them to the desired locations. CAE (computer-aided engineering) systems support conceptual design by synthesizing alternative prototypes using rule-based systems; they also support design verification through rule checking of CAD models. CAD/CAM/CAE systems are intended to increase productivity by optimizing design and production steps, and by increasing flexibility and efficiency.

Application Areas CAD/CAM/CAE systems are used in all engineering design disciplines to support product design, test, and production.

Design Problems Engineers must correctly convert real-world specifications to valid computer models, design appropriate computer tests for the model, and design computer systems to accurately and economically manufacture the final product.

Engineering Disciplines CAM has had a major effect on the work of many industrial engineers. Switching over from labor-intensive manufacturing to computer-aided manufacturing has changed and will continue to change the national and international industrial environment. Because they use CAM systems themselves, industrial engineers are heavily involved in their development. Computer and electrical engineers design the computer and control systems and write the software that drive CAM systems.

This automated integrated circuit insertion device, designed by Lockheed, is installing an integrated circuit into a circuit board for a satellite program.

8

LASERS

Light waves from a laser (that is, light amplification of electromagnetic waves by stimulated emission of radiation) have the same frequency and thus create a beam with one characteristic color. More importantly, the light waves travel in phase, forming a narrow beam that can easily be directed and focused.

Application Areas CO_2 lasers can be used to drill holes in materials such as ceramics, composite materials, and rubber. Medical uses of lasers include repairing detached retinas, sealing leaky blood vessels, vaporizing brain tumors, removing warts and cysts, and performing delicate inner-ear surgery. Lasers are used in scanning devices to scan Universal Product Codes. High-power lasers also are used in weapons and to create 3-D holograms.

Design Problems A key concern is controlling the heat and power generated by lasers so the laser does not harm the patient or the product.

Engineering Disciplines Although they were not heavily involved in the original design and development of lasers in the 1950s, engineers have developed applications for them in

Automatic analysis of DNA is performed with laser beams by Leroy Hood and Jane Sanders, biologists at California Institute of Technology. This DNA sequenator is also called the "Gene Machine."

many areas: Electrical engineers use them in optical fiber communications, civil engineers use them to perform accurate surveying, and mechanical engineers use them for precise cutting of metal parts.

9

FIBER OPTIC COMMUNICATIONS

An optical fiber is a transparent thread of glass or other optically transparent material. It can carry more information than either radio waves or electrical waves in copper telephone wires. In addition, fiber optic communication signals do not produce electromagnetic waves that cause cross-talk noise on communication lines. The first transoceanic fiber optic cable was laid in 1988 across the Atlantic. It contains four fibers that, together, can handle up to 40,000 calls at one time.

Application Areas An increasing number of communications and computer systems are converting to fiber optics due to its enormous information capacity, small size, light weight, and freedom from interference.

Design Problems A key design problem is to create cables that can withstand the stress of being buried under ground or under the ocean. Another design problem is the need to keep the various signals independent and free from outside interference.

Engineering Disciplines Mechanical engineers design the manufacturing systems that produce glass and plastic optical fibers. Electrical engineers design the transmitters, amplifiers, and receivers that, along with the fibers, carry the optical signals from source to destination.

10

GENETIC ENGINEERING

A genetically engineered product is created by splicing a gene that produces a valuable substance from one organism and placing it into another organism that will multiply itself and the foreign gene along with it. Genes are artificially recombined in a test tube, inserted into a virus or bacteria, and then inserted into a host organism in which they can multiply. Once the new organism has been created, a system has to be designed to produce and process the product in large quantities at a reasonable cost.

The first commercial product of genetic engineering was human insulin, which appeared commercially under the trade name Humulin. The molecules are produced by the genetically engineered bacteria and are then crystallized into human insulin.

Application Areas In addition to creating new drugs and vaccines, genetic engineering has been used to create bacteria that can clean up oil spills and detoxify wastes. The process is also used to create genetically altered plants that are pest and disease resistant or have certain desirable characteristics such as improved taste, shipping hardiness, or longer shelf life.

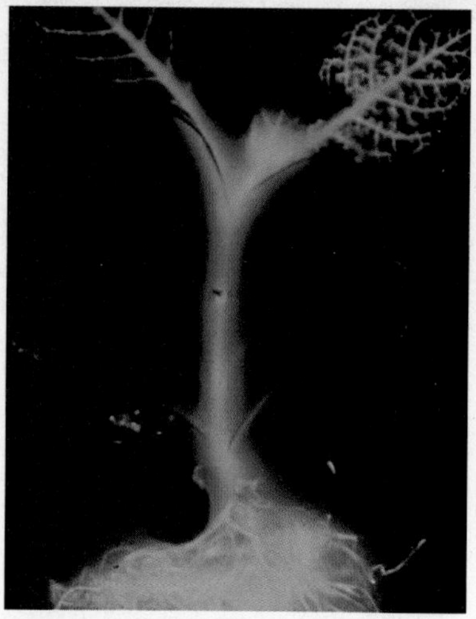

Researchers incorporated firefly gene codes for the enzyme that catalyzes the chemical reaction to release energy in the form of light, into the DNA of a tobacco plant.

Design Problems Engineers must translate the laboratory work of biologists into the large-scale commercial manufacturing systems that are robust, safe, and cost effective.

Engineering Disciplines Mechanical engineers design equipment for growing large quantities of genetically engineered organisms, chemical engineers design processes for separating out the desired end product, and environmental engineers evaluate the potential impact on the environment.

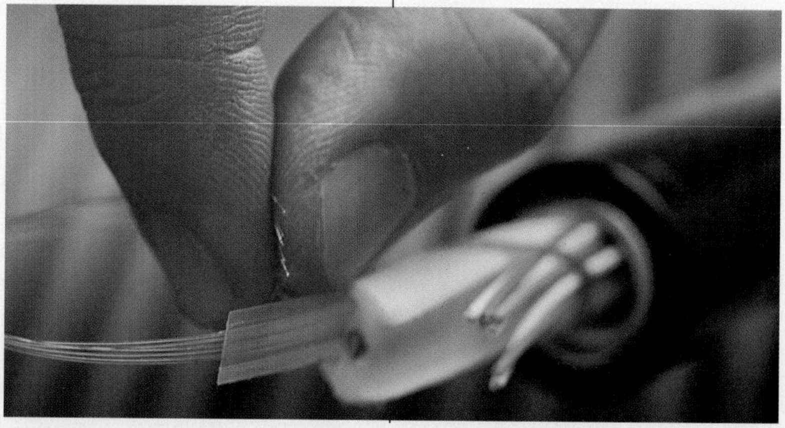

Thinner than a human hair, an optical fiber can carry more information than conventional radio waves or electrical waves.

Mathcad 6.0 for Engineers

Joe King
Department of Electrical Engineering
University of the Pacific
Stockton, California

Mathcad 6.0 for Engineers

Addison-Wesley Publishing Company, Inc.
Menlo Park, California · Reading, Massachusetts
New York · Don Mills, Ontario · Harlow, U.K. · Amsterdam
Bonn · Paris · Milan · Madrid · Sydney · Singapore · Tokyo
Seoul · Taipei · Mexico City · San Juan, Puerto Rico

Copyright ©1997 by Addison-Wesley Publishing Company, Inc.

All rights reserved. No part of this publication may be reproduced, stored in a retrieval system, or transmitted, in any form or by any means, electronic, mechanical, photocopying, recording, or any other media or embodiments now known or hereafter to become known, without the prior written permission of the publisher. Printed in the United States of America. Published simultaneously in Canada.

The applications and examples presented in this book have been included for their instructional value. They have been tested with care but are not guaranteed for any particular purpose. The publisher does not offer any warranties or representations, nor does it accept any liabilities with respect to the applications and examples.

MathSoft and Mathcad are registered trademarks of Mathsoft, Inc. Some of the product names used herein have been used for identification purposes only and may be trademarks of their respective companies.

This is a module in the *Engineer's Toolkit*, an Addison-Wesley SELECT edition. Contact your sales representative for more information.

The Engineer's Toolkit and SELECT are trademarks of Addison-Wesley Publishing Company, Inc.

Photo Credits:
Chapter 1: ©Nick Cedar
Chapter 2: ©Robert Rathe/Stock Boston
Chapter 3: ©Richard Pasley/Stock Boston
Chapter 4: ©David R. Frazier/Tony Stone Images
Chapter 5: ©Tony Stone Images

ISBN: 0-8053-6484-6

Addison-Wesley Publishing Company, Inc.
2725 Sand Hill Road
Menlo Park, CA 94025

Contents

1 Solving Problems with Mathcad

Motorcycle Grand Prix

Microprocessors deserve much of the credit for the efficiency, dependability, and performance of the motorcycles that participate in the world Grand Prix (GP) racing circuit. Among other tasks, microprocessors store data on parameters such as suspension system movement; water, oil, and tire temperature and pressure. At the end of each practice ride, a technician transfers the data to a trackside computer. Engineers then plot and analyze the data, and instruct mechanics to make appropriate adjustments to the motorcycle. In this way, engineers can coax a 500cc (0.5 liter) GP motorcycle engine into generating 200 horsepower. Engineers also design the high-strength, lightweight chassis; the high-power, lightweight engines; and on-board computer systems that help make today's level of competition possible. To achieve such goals engineers make use of specific problem-solving skills. This chapter helps develop these skills by introducing a five-step problem-solving process.

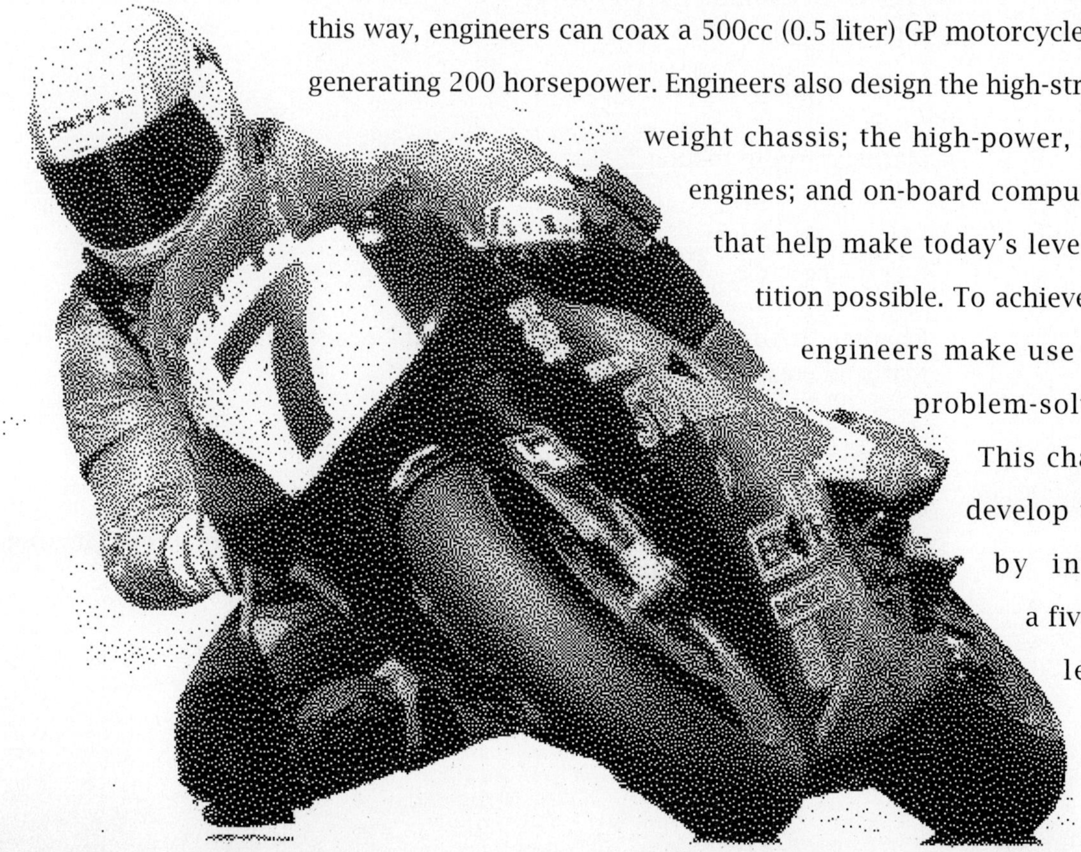

INTRODUCTION

The primary goal of this module is that you become competent at quick and efficient editing, solving, and graphing of the kinds of mathematical problems that you will likely have to solve as a lower-division engineering student. These problems range from simple arithmetic to calculus.

Two secondary goals of this module are (1) that you learn to enjoy using Mathcad so that you will want to use it to solve your problems and (2) that you know what Mathcad can do, even if you cannot yet use its advanced features. In this way, you may be motivated to go beyond this module and teach yourself how to use Mathcad's advanced features in your future work as an engineering student and practicing engineer. This module focuses on Mathcad 6.0 for Windows 95.

This chapter introduces you to problem solving by first examining the types of problems Mathcad can solve. We then present a five-step process that is useful for solving typical engineering problems and use a simple example to illustrate the use of the five-step process. We conclude with an overview of Mathcad's conventions and features.

1-1 PROBLEM SOLVING WITH MATHCAD

Mathcad is one of the most easily learned yet most powerful engineering tools you could have at your disposal. You will probably even find it fun to use. It combines the capabilities of a programming language and a spreadsheet but is much easier to learn to use.

Using Mathcad, you type equations and expressions in somewhat the same form as you would in a computer program or spreadsheet, yet they appear on the screen in the same form in which you would write them on paper. For example, were you to type within the Mathcad window (without typing (ENTER))

```
(a + b) / (c - d)
```

Mathcad would display

$$\frac{a+b}{(c-d)}$$

Chapter 2 introduces equation, text, and graph entry and shows how Mathcad easily solves equations such as

$$3.1a^2 = 83.824$$

Chapter 3 covers functions, vectors, matrices, complex numbers, random numbers, histograms, statistics, the solution of sets of equations, sorting, and iterative calculations such as

$$Xave := \frac{1}{n} \cdot \sum_{i} (X_i)$$

Chapter 4 describes how to create complex graphs, polar plots, and contour plots such as the following in just a few seconds:

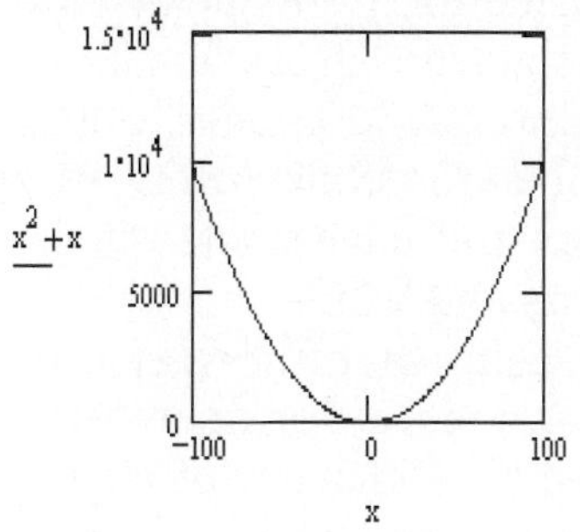

In Chapter 5, you will discover that Mathcad can even calculate differentials and integrals such as

$$\mathrm{Irms} := \sqrt{\omega \cdot \frac{\mathrm{Im}^2}{2\cdot\pi} \cdot \int_0^{2\cdot\frac{\pi}{\omega}} \cos(\omega\cdot t + \phi)^2 \, dt}$$

You can use Mathcad to solve just about any mathematical problem that would arise in engineering. All the Greek letters and operators are available at a click of the mouse. Mathcad can find everything from the largest number in a matrix to the integral of complex trigonometric functions, and much more.

Mathcad's two- and three-dimensional graphing capabilities give you, the engineer, a visual representation of your data. Mathcad can plot single functions, multiple functions, polar plots, and 3D surface plots. Graph axes can be linear or logarithmic.

Its symbolic processor gives Mathcad unprecedented capabilities beyond those of spreadsheets. The *symbolic processor* is a program that allows Mathcad to simplify complex expressions, factor expressions, determine the value of a variable that will solve an equation, and differentiate or integrate an expression in terms of any variable, among other operations. In milliseconds, Mathcad can complete a mathematical operation that would be difficult or impossible for an engineer to do by hand.

Mathcad allows you to add text to your equations and graphs and will then print your document exactly as it appears on the screen. The result is a nice, neat record of your work that you can proudly insert into any engineering report. Furthermore, since Mathcad is a Windows application, you can easily copy all of your work into any Windows-based word processing program, such as Microsoft Word or WordPerfect for Windows (95). In preparing early drafts of this text, we copied the preceding example equations and the contour plot into this Word document from a Mathcad document.

1-2 THE FIVE-STEP PROBLEM-SOLVING PROCESS

Problem solving is what engineering is all about. For centuries, engineers have applied scientific principles to the solution of real-world problems. For example, engineers in ancient Rome solved transportation problems by building highways and bridges. Contemporary engineers have addressed the problem of automobile safety by developing air bags and antilock braking systems and have made our lives more convenient by designing microwave ovens and VCRs.

However, it seems that for every problem engineers solve, society asks them to solve two more. Fortunately, to this end, computers have significantly increased the efficiency of engineers. Computers never forget what they are taught, never get bored or tired, and never complain about working 24 hours a day. They can collect data, analyze data, and even propose problem solutions based on analyzed data.

Whether or not an engineer uses a computer, he or she must approach each problem methodically, using an *algorithm.* An algorithm is a step-by-step procedure by which one arrives at a solution to a problem. For example, a cooking recipe is an algorithm.

Typically, when an engineer uses a computer, the algorithm must be described to the computer using a programming language such as C or C++. Mathcad does have built-in programming capabilities (see the Mathcad features list later in this chapter). However, although you may choose to customize its features so that they function more to your liking, Mathcad does not require you to do any programming.

Once you create a problem solution with Mathcad, it appears on the computer screen looking just as it would on paper. Summation, integration, and differentiation symbols, for example, are perfectly drawn. You can then save the solution on a disk and use it as often as you like. You can easily create iterative solutions and apply them to problems that involve millions of iterations and hours of processing.

Whatever the problem, the engineering process can be reduced to a five-step problem-solving process. A general process is shown on the left and the process you will use to solve problems with Mathcad is shown on the right. The two processes are nearly identical; there are only minor semantic differences.

General Problem-Solving Process	**Mathcad Problem-Solving Process**
1. Define the problem.	1. State the problem clearly.
2. Gather information.	2. Gather information.
3. Generate and evaluate potential solutions.	3. Generate and evaluate potential solutions.
4. Refine and implement a solution.	4. Refine and implement a solution.
5. Verify and test the solution.	5. Verify the solution through testing.

1-3 APPLYING THE FIVE-STEP PROBLEM-SOLVING PROCESS

Let's demonstrate the five-step problem-solving process by taking a closer look at how the performance of a racing motorcycle might be improved.

Step 1: Define the Problem Clearly. Exactly what is it that must be accomplished? What are the constraints? In this example, the problem is keeping the rear tire of a racing motorcycle on the ground during a race—that is, the tire should not bounce off the ground over the bumpy parts of the course, so that the motorcycle consistently uses the engine's power for propulsion.

Step 2: Gather Information. We mount movement sensors on the motorcycle's rear suspension system so that they measure the movement of the rear tire relative to the ground. A microprocessor-based system reads the outputs of the sensors once each millisecond and delivers that data to a trackside computer after each practice ride.

Step 3: Generate and Evaluate Potential Solutions. Generating and evaluating potential solutions often involves the collection, plotting, and evaluation of data. The trackside computer saves the following sensor data in a data file called shock.dat:

0.0,0.1,2.3,0.0,0.0,0.0,0.3,2.3,0.0,0.0,0.2,1.9,0.0,0.0,0.6,2.0,0.1,0.0,0.
0,0.0,0.0

Each data point represents the distance (in centimeters) that the rear tire lifted off the ground. Data were collected over a 20-millisecond period.

Mathcad reads the contents of shock.dat into a vector S and plots the data against time. The graph (below) shows that the rear tire left the ground four times during the 20-millisecond period. The Mathcad function max is used to determine the maximum distance the rear tire bounces: 0.023 meter.

$i := 0 .. 20$

$S_i := \mathrm{READ}(\mathrm{shock}) \cdot \mathrm{cm}$

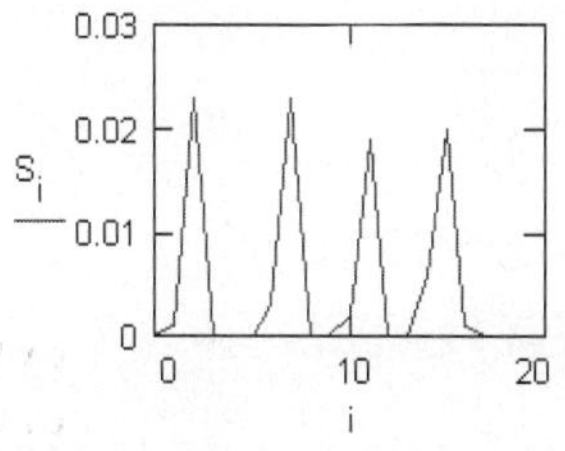

$\max(S) = 0.023 \cdot \mathrm{m}$

From this information, an engineer might propose that the rear shock compression damping be decreased. An alternative is to decrease the amount of air in the tire, which would make the tire more shock absorbent. However, this would cause the tire to deform as the racer leans deeply into turns, so this solution is not acceptable.

Step 4: Refine and Implement a Solution. We instruct the mechanic to adjust the rear shock compression damping, decreasing it. Engineering solutions may be as simple as a shock adjustment or as complex as the construction of a skyscraper. A typical engineering solution will involve a complex sequence of operations. When this is the case, the engineer typically uses a *top-down design* process, which involves looking at the problem as a whole and then breaking it down into smaller problems. This process, also called *stepwise refinement*, continues until it results in problems that can be readily defined and solved.

In our present example, our ultimate goal is to win the race; the problem is how to maximize the chances of winning. We can break down this problem into the subgoals of improving rider performance and improving motorcycle performance. We can then break down the goal of improving motorcycle performance—the subgoal under the engineer's control—into the more specific goals of increasing power and improving the handling characteristics. Better handling can be broken down further into achieving faster corner speeds and keeping the tires on the ground.

Step 5: Verify the Solution Through Testing. As when evaluating potential problem solutions, verifying the final solution generally involves the collection, plotting, and analysis of data. We take the motorcycle onto the track and ride it at a race pace. We again collect the relevant data, put it into the computer, plot it, analyze it, and check the results. Is the rear tire staying on the ground? If not, we try another solution.

As a future engineer, you must learn to make such a process second nature in solving engineering problems. In each of the remaining chapters, we offer an application of this five-step process. The discipline and the domain of each application problem follow:

Applications	**Across the Disciplines**	
		Chapter
Signal propagation	Electrical engineering	2
Computer networks	Computer engineering	3
Topographic surveys	Civil engineering	4
Composite materials	Mechanical engineering	5

1-4 MATHCAD FEATURES

In this section, we list the more useful of Mathcad 6.0's many features. Note that Mathcad 6.0 is available in standard and Plus versions. All of the features listed below, except for programming, are included in both versions. Considering the limited space in this module, however, we will not be able to elaborate on all of these features. In any event, many of Mathcad's capabilities are better left to upper-division and even graduate-level engineering students. See your user's manual or Mathcad's help features for more information.

User Interface

- Context-sensitive online help
- Example worksheets available as QuickSheets and Electronic Books
- Free-form, scratchpad-type data entry
- Mixed text, mathematics, and graphs throughout the document
- Error messages that flag specific parts of any errant expression or equation
- Mathematical rule checking performed by an expert system
- Cut and paste capability for all text, mathematics, and graphs
- Easily entered Greek letters and operators
- Full Windows and Windows 95 compatibility, providing the ability to resize and move windows, open multiple windows, and cut and paste between Mathcad documents and documents from other Windows programs such as word processing programs

Computations

- Fifteen decimal digits of accuracy
- Symbolic processing that provides exact answers (for example, 2/3 vs. 0.666...) for operations such as expression factoring and determining derivative and integral expressions
- Built-in measurement units and dimension checking
- Equation solving that includes derivative and integral computations
- Vector and matrix operations that include matrix multiplication, inverse, transpose, determinant, and dot and cross products
- Complex numbers, complex variables, and complex functions
- Summation and product operations
- Octal, hexadecimal, and decimal numbers
- Trigonometric, hyperbolic, exponential, and Bessel functions
- Statistical functions that include mean, standard deviation, linear regression, gamma function, erf, and cumulative normal distribution
- Cubic spline curve fitting
- One- and two-dimensional Fast Fourier Transforms (FFT)

Graphing

- A variety of graph types, including rectangular (x-y) plots, polar plots, surface plots, contour plots, vector field plots, three-dimensional bar charts, and scatter plots
- Linear and logarithmic graph axes
- One or more traces per graph, identified with legends
- Graphics that can be imported and exported
- Plot animation for viewing dynamic processes

Text

- Text anywhere in the document with automatic word wrap and paragraph formatting
- Spell checking
- Multiple fonts, sizes, and styles in any text region
- Text printed in WYSIWYG (what you see is what you get)

Input/Output

- Import and export ASCII data files, including columns of numbers from spreadsheets
- Compatible with Microsoft Mail and cc:Mail
- Open, and establish hypertext links to, worksheets anywhere on the Internet
- Export worksheets directly to Microsoft Word documents
- OLE and DDE support

Programming (Mathcad PLUS 6.0 only)

- Conditional branching
- Looping constructs
- Recursive definitions

SUMMARY

This chapter introduced you to Mathcad 6.0, starting by briefly describing the program's problem-solving capabilities and ending with a detailed list of its features. The chapter presented a five-step problem-solving process and an application of the process. The five steps are as follows:

1. Define the problem.
2. Gather information.
3. Generate and evaluate potential solutions.
4. Refine and implement a solution.
5. Verify and test the solution.

Key Words

algorithm	symbolic processor
stepwise refinement	top-down design

2

Introduction to Mathcad 6.0

Telephone Transmission

The engineers who develop and maintain the worldwide telephone system often refer to it as POTS (plain old telephone service). Today, however, little is plain about it. POTS is a complex global network of telephones, information-carrying media, and switching equipment. It digitizes voices near their origin before placing them in the system and then converts them back to analog near their destinations. So, traveling over the bulk of the system are simply billions of bits representing binary pieces of millions of voices. Copper wire, optical fiber, earth's atmosphere, and space are the primary media that carry this bit traffic all over the world. In space, satellites transfer digitized voices from one point on the globe to another. The satellites used by POTS sit in orbits 22,300 miles above the earth, creating a propagation delay—pauses between words being said by one party and heard by the other—of about one-fourth second. In this chapter, you will see how Mathcad calculates this propagation delay.

INTRODUCTION

In this chapter, you will learn the basics of Mathcad 6.0. This version of Mathcad is essentially the same as Mathcad 5.0, but it does have a few new features, such as plot enhancements, more kinds of plots, more statistical functions, programmability, QuickSheets, locked regions, email capabilities, Internet access, and animation. Most of the differences between Mathcad 6.0 and Mathcad 5.0 deal with advanced features that this module does not cover. In particular, all the essential commands and techniques that we describe are applicable to versions 5.0 and 4.0 as well. Although the appearance of version 6.0 differs considerably from earlier versions and some commands in the pull-down menus have been moved or renamed, users of earlier versions should find most discrepancies in the Mathcad descriptions in this module minor.

Section 2-1 of this chapter begins with a description of the conventions used in this module. The section explains how to run Mathcad and describes the Mathcad window. The section closes with a look at Mathcad's help features and the composition of Mathcad documents. The creation of documents is the first topic of Section 2-2. A brief explanation of how to enter equations, text, and graphs follows. The section also presents a description of how to save and print documents. Section 2-3 covers the editing of documents. This section includes an explanation of how to use variables in your equations and how to use Mathcad's predefined constants, such as π and e. A brief discussion of error messages follows. The chapter concludes with an application problem, in which the five-step problem-solving process is used.

Virtually all the subjects introduced in this chapter are covered in more detail in the remaining chapters of the module. In the meantime, be sure you have Mathcad running on your computer so that you can work through the examples and Try It! exercises. You will be surprised at how quickly you learn to use Mathcad if you try every feature as you go through this material.

2-1 GETTING STARTED

Module Conventions

This module uses the following conventions:

Italics represent scalar variable names, function names, and key words.

A `monospaced font` represents characters you should type and error and other on-screen messages.

Bold represents menu titles and selections, vectors and matrices.

Special keys are denoted using appropriate symbols. For example, (↑), (↓), (←), and (→) denote the arrow keys. The function keys are shown as (F1), (F2), and so on; the Escape key as (ESC); the Delete key as (DEL); the Backspace key as (BKSP); and the Tab key as (TAB). The Shift, Control, Alt, and Enter keys are shown as (SHFT), (CTRL), (ALT), and (ENTER) respectively.

When two keys are shown together with a plus (+) sign between them, as in (CTRL)+a, you should press the first key, hold it down and then press the second key. When a sequence of keys separated by commas is shown, as in ALT,f,x, you should press and release each key in the order shown.

All mouse clicks refer to the left mouse button. When told to select an option under a menu, you should click the mouse on the indicated menu name on the *menu bar* and then click on the indicated selection in the pull-down menu that appears. The menu bar is the second row from the top of the screen. For example, if told to quit Mathcad by selecting **Exit** under the **File** menu, you would click the mouse button on the **File** menu name and then click on the **Exit** selection in the menu that appears.

You can perform nearly all operations in Mathcad using *shortcut key sequences.* For example, you can quit Mathcad simply by pressing the key sequence: ALT,f,x. Generally, you will be instructed to use the mouse to perform Mathcad operations since "select **Exit** under the **File** menu" more clearly indicates what needs to be done than does "press ALT,f,x." However, we will occasionally remind you of shortcut key sequences that you may prefer to use. In any event, Mathcad indicates all shortcut keys by underlining them in the names of the menus and the selections under the menus. All shortcut key sequences begin by pressing and releasing the ALT key.

Running Mathcad

To run Mathcad, you must first run Windows or Windows 95. If you are running Windows 95, Windows came up automatically when you turned on your computer. If not and you're running Windows 3.1, at the DOS prompt enter: winENTER. Once you are up and running in Windows, either a Mathcad icon, labeled Mathcad 6.0, or a Mathcad Program Group icon (a square) should be present.

If a Mathcad Program Group icon is present, you must double-click the mouse on it to gain access to the Mathcad icon. Once you see a Mathcad icon, double-click on it to run Mathcad. You should run Mathcad now and familiarize yourself with it as we describe the program.

As you get started, always keep in mind that the **Help** menu offers various help options. Particularly useful is the **Index** option. As with most Windows-based programs, pressing the F1 key will call up the **Help** menu index. The **Search** selection under the **Index** option allows you to quickly access help on any Mathcad subject. When you select **Search**, a dialog box appears that asks you to enter the subject on which you want help. For example, explanations of all error messages can be listed by entering "error messages" in this dialog box. To get immediate help on an error message, click once on the message and then press F1. To get context-sensitive help at other times, you can, for example, click on any mathematical operator (such as the integration operator) and press F1. To remove the **Help** window, double-click on the button in the window's extreme upper left-hand corner.

The Mathcad Window

When Mathcad first starts, its window may be somewhat smaller than full screen. If so, click the mouse on the small up arrow (or the small, square box if you are using Windows 95) in the extreme upper right-hand corner of the window. The Mathcad window will expand to full screen.

Figure 2-1 displays the Mathcad document window. At the top is the *title bar,* where the words *Mathcad (or Mathcad PLUS)* are followed by the name of the Mathcad document you are editing. Its default name is Untitled:1, but you can give it any legal filename when saving it by selecting **Save As** under the **File** menu. In Section 2-3, we will discuss file saving in more detail. At the left end of the title bar is a Mathcad icon/button; double-clicking on this button (or the X button in the upper right-hand corner if you are using Windows 95) will quit Mathcad.

Figure 2-1
The Mathcad Document Window

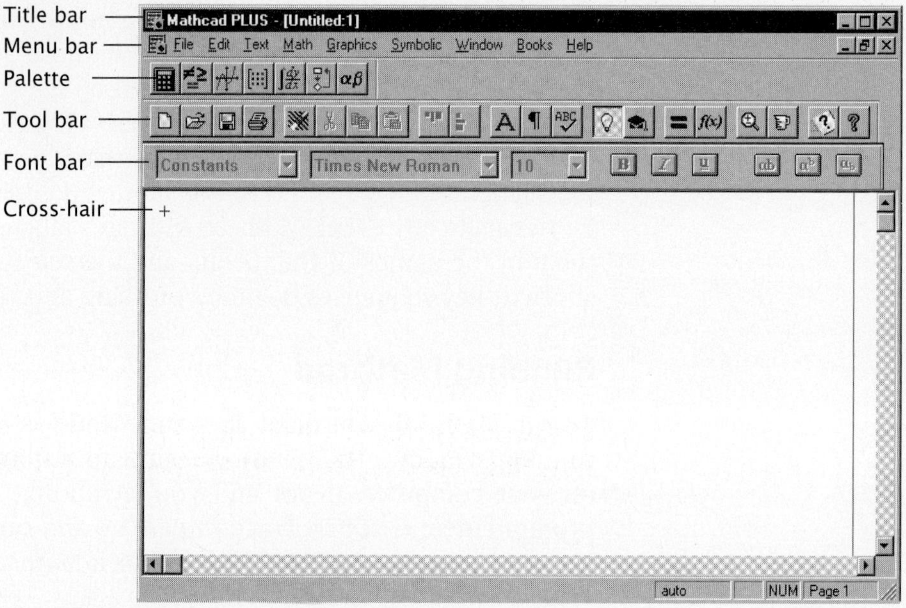

As we indicated earlier, the second row from the top of the Mathcad window is the menu bar, with **File** at the left end and **Help** at the right end. To pull down a menu, you can either click on it with the mouse or press the ALT key followed by the underlined letter in the name of the desired menu. The underlined letter is a shortcut key. You can either hold down the ALT key as you press a shortcut key or release it before you press the shortcut key. Many of the menu selections also can be activated using the CTRL key in combination with another key. For now, you can examine the available menus by pulling down each one. Much of what you see will be familiar to you if you are accustomed to using Windows applications. We will describe most of the menus in more detail later, but for now they should be somewhat self-explanatory. At the left end of the menu bar is another Mathcad 6.0 icon (in Windows 95); double-clicking on this button will close the present Mathcad document.

The third row from the top of the Mathcad window is the *palette.* By clicking on the buttons on the palette, you can easily insert arithmetic operators, evaluation and Boolean operators, graphs, vectors and matrices, calculus operators, programming templates, and Greek letters. If you leave the mouse arrow momentarily over any button on the palette, or any other button on the screen, Mathcad will display a concise description of the button's function.

The fourth row from the top of the Mathcad window, called the *toolbar*, contains buttons on which you can click in order to quickly perform some of Mathcad's most commonly used operations. Most of these buttons are similar to those displayed in most Windows-based programs. The operations they initiate include retrieving, saving, and printing documents, spell checking, function and units selection, and accessing help. You can also access all the toolbar operations using the pull-down menus.

To learn what each toolbar button does, press SHFT+F1. A large question mark will appear adjacent to the mouse arrow. Click on any button to get an explanation of the function of that button. Press ESC to exit this help feature.

Below the toolbar is the *font bar*. The font bar contains pull-down menus that allow you to select text styles, fonts, and sizes, and buttons that allow you to select bold, italics, underlining, subscripts, and superscripts.

Although the toolbar and the font bar offer the user many convenient shortcuts, for consistency (since different versions of Mathcad vary in this regard), in this module all operations will be specified via the pull-down menus. Therefore, we do not mention the toolbar and font bar again. If you prefer not to use the palette, tool, and font bars (preferring to use either shortcut keys or the pull-down menus) and want more room for your work, you can remove them from your window by selecting **Hide Palette**, **Hide Toolbar**, or **Hide Font Bar** under the **Window** menu.

Help

Mathcad 6.0 offers as much help as you could want in getting started on any task, simple or complex. As mentioned above, the **Help** menu and the F1 key call up the usual kind of help found in most Windows programs. However, Mathcad also offers extensive example problems in all areas of computation. These examples are grouped into Mathcad's QuickSheets and Electronic Books.

QuickSheets offer specialized help in areas from arithmetic to calculus. QuickSheets are accessed either by clicking on the toolbar button with the light bulb on it or by selecting **QuickSheets...** under the **Help** menu. The **QuickSheets** menu then offers you help in many mathematical areas, as well as access to the World Wide Web and the Mathsoft (creators of Mathcad) web site. Copies of the contents of any QuickSheet can be dragged easily into user worksheets; just select what you want by dragging the mouse across it and then use the mouse to drag the selected area into the worksheet. Figure 2-2 displays the **QuickSheets** table of contents.

Figure 2-2
The Mathcad Quicksheets Menu

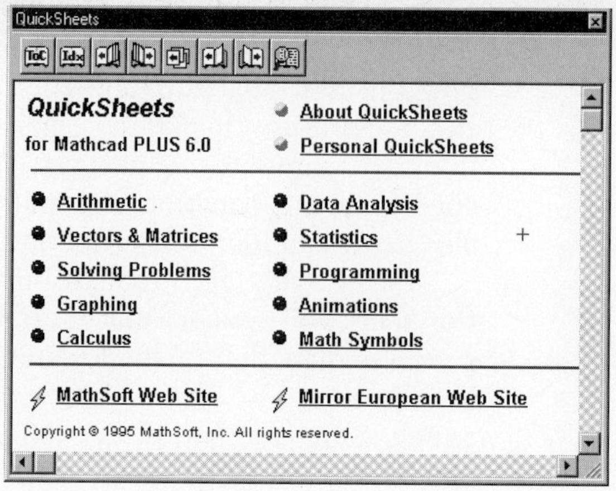

Mathcad's *Electronic Books* offer graphical examples in many specific areas of engineering. Each Book is a collection of up to several hundred Mathcad worksheets containing text, pictures, formulas, and data. Each worksheet can be edited and any of its contents copied into any user worksheet. Example worksheets include problems in electromagnetics and properties of materials. To access Electronic Books, select **Open Book...** under the **Books** menu, double click on the Deskref.hbk icon, and double click on Table of Contents. Figure 2-3 shows an Electronic Book problem in which the capacitance of a cylindrical capacitor is determined.

Figure 2-3
A Mathcad Book Example

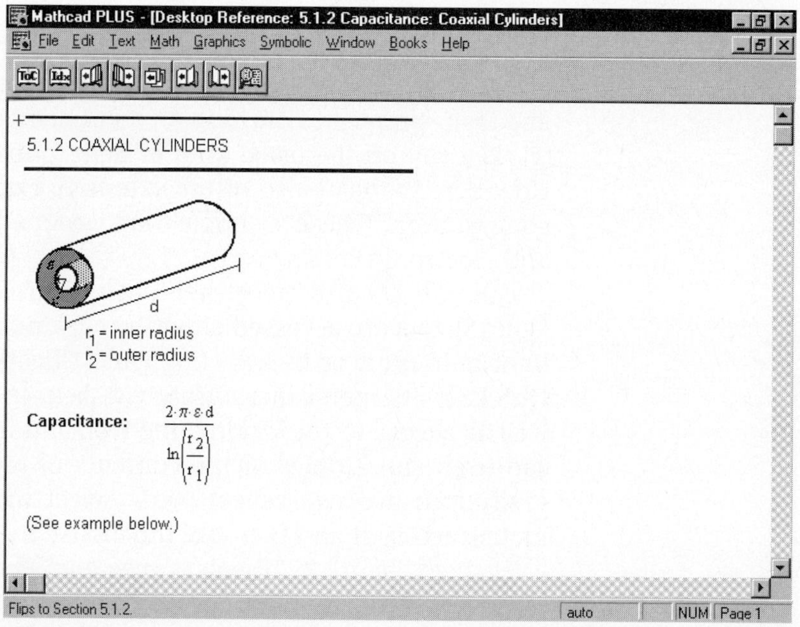

Whenever you get stuck when trying to solve any problem, consider turning to the QuickSheets and Electronic Books. The QuickSheets will be helpful in virtually all areas of mathematics; the Electronic Books will be helpful in specific areas of science and engineering.

Mathcad Documents

As you start each Mathcad session, there should be a red *crosshair* in the upper left-hand corner of an otherwise blank Mathcad document. This crosshair indicates that Mathcad is ready for you to enter new work. If it is not present, select **New** under the **File** menu. You can move the crosshair anywhere in the document by pressing the arrow keys or by placing the mouse cursor (an arrow) at the desired location and then clicking the mouse button. When you begin to type something the crosshair disappears, replaced by the character you typed, followed by a vertical blue bar. At this point, you have created a *region* within the document. A Mathcad document consists of a collection of regions.

There are three types of regions: *equation, text,* and *graph.* The default region type is equation. If you simply begin typing text when the crosshair is present, Mathcad will assume you are entering not text, but a variable name as part of an equation. To create a text region, you must first type a double quote ("). Whatever you enter subsequently will be recognized as text. To exit a text region, you must click the mouse outside the region. To create graphs and plots, you select a graph type under the **Graphics** menu.

This chapter introduces you to the creation and editing of equation, text, and graph regions. Chapter 3 goes into much more detail on the creation, editing, and use of equation regions. Chapter 4 goes into much more detail on the creation and editing of graphs and plots.

An equation region can contain only one equation. Likewise, a graph region can include only one graph or plot. However, a text region can contain as little or as much text as you like.

A document can contain as many of each type of region as you like. Mathcad maintains an invisible rectangle around each region so that you can designate any given region for operations such as editing, moving, cutting, and pasting. You can make these rectangles visible as dashed line boxes by selecting **Regions** and then **Select All** under the **Edit** menu. Another method for making the regions visible is to hold down the mouse button in one corner of the document, move the mouse cursor to the opposite corner of the window, and then release the mouse button. All regions should then have their perimeters defined by dashed lines.

2-2 CREATING DOCUMENTS

Entering an Equation

Your first Mathcad session begins with the creation of an equation region. First note, however, that you can clear the entire document window by selecting **Regions**, then **Select All** under the **Edit** menu and then pressing the DEL key. Alternatively, you can use the **File** menu to close the present document without saving it, then open a new document.

A quick way to select all regions is simply to drag the mouse across the document window as described earlier. Pressing DEL will then clear the document.

When you clear a document or open a new one, the red crosshair will be the only visible item in the document. If you have been trying out Mathcad before arriving at this point, clear the document window now.

Whenever the red crosshair is present, Mathcad is ready to create a new region. When you begin typing, an *insertion point,* appearing as a vertical line, replaces the crosshair. An insertion point will also appear if you click the mouse on any variable names, numbers, or text you have entered. Within text regions, the insertion point is red; in equation and graph regions, it is blue. Whenever the insertion point is present, you can insert characters or, using BKSP or DEL, delete them. Clicking the mouse outside all existing regions will cause the red crosshair to reappear. The red crosshair and insertion point are editing tools, one of the subjects of Section 2-4.

While entering equations, a small black box (■) often appears to the right of the operator that was just entered. This is a *placeholder* and indicates that Mathcad is waiting for you to enter something it needs. For example, when you enter a plus sign, a placeholder will indicate that a number to be added must now be entered.

Whenever you press the equals sign (=) after entering a complete expression, Mathcad will compute and show the result, followed by a placeholder for units such as meters or seconds. Although you are not required to enter units, they do enhance the display of your computed results. Mathcad will even make sure you use your units consistently and give you error messages if you do not. The use of units is covered in Chapter 3. For now, ignore placeholders for units; they disappear when you press the ENTER key.

If, during these first few steps in learning Mathcad, you make a misstep and do not yet know how to correct it, try selecting **Undo Last Edit** under the **Edit** menu. As a last resort, simply clear the screen and start over. (We cover editing in detail in Section 2-4.)

Occasionally, an error message or overlapping regions will cause a portion of the screen image to go blank or become otherwise corrupted. If this occurs, select **Refresh** under the **Window** menu to restore the full screen. Pressing CTRL+r will achieve the same effect. You also may have to move the regions farther apart from each other because the regions grow as you enter text or equations, and one may overlap another. You can separate regions either by selecting **Regions** and then **Separate** under the **Edit** menu or by simply pressing CTRL+s.

EXAMPLE 2-1

Entering an Equation

Determine the value of x when $x = 47-5(6.5)$.

SOLUTION

Before beginning, make sure the cursor, a red crosshair, appears in the Mathcad window. It should be in the upper left-hand corner. If not, click the mouse anywhere in the window where you want to begin; the red crosshair should appear. If it does not, use the **File** menu to open a new document.

Type: `47-5*6.5=`

With the press of the 4 key, you created an equation region. A blue insertion point replaced the red crosshair. When you typed the minus sign, a placeholder appeared, indicating that Mathcad is waiting for a number that will be subtracted from 47. The equals sign (=) is called the *evaluation operator* in Mathcad. When you press the equals sign, Mathcad evaluates the expression or variable to the left of the equals sign and another placeholder appears. This placeholder indicates that Mathcad is waiting for you to enter (optional) units.

Press: ENTER

The units placeholder disappears, and the red crosshair reappears below the equation.

On the screen, you should now see

$$47 - 5 \cdot 6.5 = 14.5$$

Mathcad computes and displays the result of any calculation as soon as you press the equals sign (=) unless you instruct it to delay screen updates by turning off the **Automatic Mode** under the **Math** menu. When Automatic Mode is off, you must press the F9 function key to produce the results of any entered equations.

Leave your work on the screen or save it in a file called example.mcd. To save a document that has not been previously saved, select **Save As** under the **File** menu. Subsequent saves can be accomplished by selecting **Save** under the **File** menu. To retrieve a saved document so that you can add work to it, select **Open** under the **File** menu. We will explain this process more completely in Section 2-3.

Try It Find the solution to the equation $x = 16 - 6/10.5$. Leave this work on the screen or save it in a file called tryit.mcd.

Mathcad uses the same mathematical operator precedence used by most calculators and computer languages. Therefore, in the preceding example, Mathcad performed the division before the subtraction. You can always use parentheses to force whatever sequence of operations you desire.

Entering Text

As we noted earlier, you create a text region when a red crosshair is present in the document and you press a double quote ("). Once you press the double quote, a box that is defined by a black line appears to contain the text. The box, and therefore the text region, grows as you type, wrapping around to the next line only when you reach the right margin or press (ENTER).

EXAMPLE 2-2

Entering Text

To the work you entered in Example 2-1, add the following text: `An Example Mathcad Equation`. If you saved your work from the previous example, you can retrieve it by selecting **Open** under the **File** menu. Then select the correct filename, for example, example.mcd, in the directory that appears.

SOLUTION

Click the mouse to the right of the equation you entered in Example 2-1, moving the crosshair to this location.

Press the double quotes key `(")` to create a text region.

A black box appears, delimiting the text region.

Type: `An Example` (ENTER) `Mathcad Equation`

Click outside the text region.

You should now see the following on your screen:

$47 - 5 \cdot 6.5 = 14.5$ An Example
Mathcad Equation

Note how easily you can move the insertion point within the text region and how easily you can edit the text.

Try It

To the work you entered in the previous Try It! exercise, add the following text: `A Simple Mathcad Equation`. If you saved your work from the previous Try It!, you can retrieve it by selecting **Open** under the **File** menu. Then select the correct filename, i.e., tryit.mcd, in the directory that appears.

Sometimes you want to set the size of a text region to be smaller than the width of the document and have the text automatically wrap within that region. To do this, type until the text region is the width you desire, then type a space and press (CTRL)+(ENTER). All other lines will then stay within the set width. To exit the region, simply click outside it.

For relatively large amounts of text, Mathcad offers the *text band.* Text bands are text regions that stretch from the left margin to the right margin and can have any height. To create a text band, start by clicking where you want the text to begin. There must be no other regions of any type between the left and right margins of the document on this line. Select **Create Text Paragraph** under the **Text** menu. Begin typing into the text band. As you create more lines of text, Mathcad moves down any regions below the text band so that the text band does not overlap them. (See the discussion of overlapping regions in "Moving Regions" in Section 2-4.)

Entering a Graph

Although we will cover the creation of graphs in detail in Chapter 4, for now you should create one just for the experience. Example 2-3 walks you through the graphing of a parabolic function.

EXAMPLE 2-3

Entering a Graph

Create a graph that plots x^2+x as x varies from -100 to +100 in increments of 1.

SOLUTION

Clear the document window.

When plotting a graph, you must first define the range of the independent variable, that is, the range of values along the X-axis. To label the independent variable as x and set its range from -100 to +100,

type: `x:-100;100`(ENTER)

Pressing the colon (:) lets Mathcad know you want to give x a value or range of values. In Mathcad, the colon is the *assignment operator.* It is used to assign a constant, variable, expression, or function name to the left of the colon a value that is to the right of the colon, as in a:cos(3*t)+5. On the screen, Mathcad inserts an equals sign after the colon to indicate the fact that you, and not Mathcad, have provided the value to the right:

$a := \cos(3t)+5$

Pressing the semicolon (;) lets Mathcad know that you are entering the initial and final numbers of a range of numbers. The line you just entered appears on the screen as

$x := -100..100$

The default stepsize is 1. To set a different stepsize, follow the initial value with a comma (,), enter the next value in the sequence (do not enter the stepsize), and then enter the semicolon and the final value. For example, to set a stepsize of 0.1,

type: `x:-100, -99.9;100`(ENTER)

Now, to create the graph,

select **Create X-Y Plot** under the **Graphics** menu.

A box should appear with placeholders along the bottom and left-hand sides:

The bottom side of the box is the X-axis; the left side is the Y-axis. A blue selection box (discussed in Section 2-4) appears around the placeholder below the center of the X-axis because Mathcad expects you to enter the independent variable first.

> To designate an independent variable, type: x
>
> Click on the placeholder to the left of the center of the Y-axis.
>
> Type: x^2(↑)(↑)+x(ENTER)

Pressing (↑) twice enclosed x^2 in a selection box so that the addition operation (+) applied to x^2 and not just to the 2.

You should now see the following on your screen:

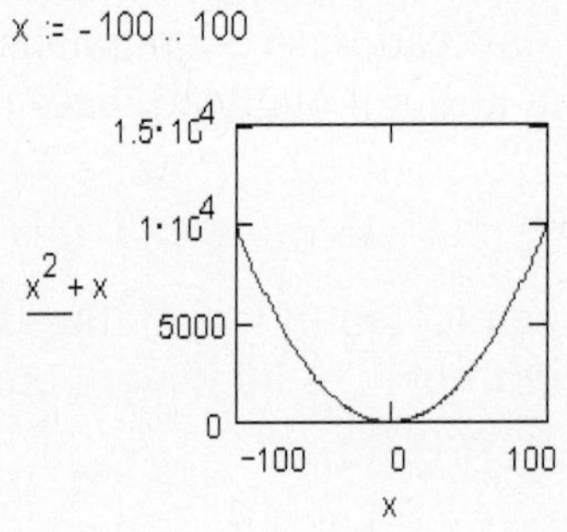

Mathcad automatically sets the range of the Y-axis for you. The line under x^2+x to the left of the graph indicates the line type of the function plotted, so it can be differentiated from other plotted functions. If more than one function is plotted, each function definition is listed to the left of the Y-axis, underlined with a unique line type. For example, the first function always is plotted with a solid line, while the second is plotted with a dashed line.

2-3 DOCUMENT MANAGEMENT

Saving and Retrieving Documents

If you wish to leave Mathcad and resume your work at a later time, you must save your document to a disk. As we stated earlier, to save a document that has not been previously saved, select **Save As** under the **File** menu. A dialog box will appear, asking you to supply a name for the document, using mcd as the default file extension. You can specify the disk drive and/or directory using a pull-down menu within the dialog box or by simply entering the information before the filename, as in b:\mcd-docs\mydoc.mcd. Select **Save** under the **File** menu to perform subsequent saves. Caution: When you save a file, your previous work will be overwritten. Furthermore, if your work is important, you should save it often to prevent its loss in case your computer loses power or crashes.

To retrieve a saved document, select **Open** under the **File** menu. A dialog box will appear listing available Mathcad documents. You may double-click on the name of one of the listed documents or enter a filename. You can specify the disk drive or directory in the same manner as when saving a document the first time.

You can export Mathcad worksheets directly to Microsoft Word documents and to Lotus Notes databases. You can also send worksheets to colleagues via email (supported systems include cc:Mail, Microsoft Mail, and NotesMail). Worksheets can be linked over the World Wide Web or on a local Web server (Mathcad supports standard Microsoft-compatible Internet access methods).

Printing Documents

To print all or part of a document, select **Print** under the **File** menu. The dialog box that appears will depend on the printer that was installed under Windows. Note that you can create Mathcad documents that are not only many pages long, but also wider than the width of typical printer paper. When a document contains multiple pages, Mathcad prints it just as any word processing document would be printed. When the document is wider than can be fit between the margins of the printed page, Mathcad will print the material to the left of the right margin and then print the material to the right of the right margin. Therefore, Mathcad will print a five-page document on ten pages if there is any material to the right of the margin on page 5.

You can use the Print dialog box to set print parameters such as page size, margins, and print quality. If you want help on printing, press F1. You can then browse the index or select **Search** on the command bar of the Help window and then enter the subject on which you desire information.

2-4 EDITING DOCUMENTS

Editing Tools

Mathcad provides three editing tools: a crosshair, an insertion point, and a selection box. Only one of these tools is active at any given time during a Mathcad session.

As we noted earlier, the red crosshair is the first thing you see at the start of a Mathcad session. You use it to select a position for any new region, whether equation, text, or graph. To place the red crosshair anywhere in your document, click the mouse at the desired location.

You can activate an insertion point within any of the three region types. An insertion point will appear if you click the mouse in a text region or on a variable name, function name, or user-defined number in an equation or graph region. You can accomplish the same thing by moving the red crosshair onto the item to be edited and then pressing the (ENTER) key.

Once you place an insertion point within text or within a variable name or number, you can use the (BKSP) key to delete characters to the left and the (DEL) key to eliminate characters to the right. You can enter new characters by typing them at the insertion point. While editing, you can move the insertion point to the left or right by using the (←) or (→) key.

The third editing tool is the *selection box*. You use this tool, a blue box, to insert and delete mathematical operators and to delete parts or all of an equation or graph region. By default, the upper right-hand corner is cut off the selection box, and insertion of an operator occurs to the right. Pressing the (INS) key moves the cut-off corner to the left and turns the selection box red. Insertion now takes place on the left of the selection box.

For example,

type: `x^2`(↑)(↑)`+1`

Now, to define a function by making *f*(*x*) equal to that expression, press (↑) until the selection box encloses the entire expression. Press (INS) to move the cut-off corner to the left.

Now type: `:f(x)`(ENTER)

The screen should now show

$$f(x) := x^2 + 1$$

You have now defined a function. Note that *f*(*x*) is often used to represent a general function, the name of which is *f* and the value of which depends on *x*. Any other text-based name for the function could have used, for example, *g*(*x*) or *position*(*x*). To find out what x^2+1 equals when x equals 1.1,

type: `f(1.1)=`(ENTER)

The creation of user-defined functions is described in more detail in Chapter 3.

When editing an equation or graph region, pressing the (↑) key will convert an insertion point into a selection box. Pressing the (↑) key again makes the selection box larger, enclosing more of any equation; pressing the (↓) key makes it smaller, enclosing less of an equation. Pressing the spacebar has a similar, but not always the same, effect as does pressing the (↑) key.

You can manipulate whatever you have enclosed in a selection box in several different ways. We describe some of the most common uses of the selection box next.

You can cut or copy the box contents to the Windows Clipboard by selecting **Cut** or **Copy** under the **Edit** menu. You can then **Paste** the contents of the Clipboard to the same or other Mathcad document or to any other Windows application, including word processing programs. If the selection box encloses a single variable or number, you can press the (DEL) key to delete that variable or number.

If a selection box encloses a mathematical expression, you can press the (DEL) key to delete the *outermost operator* in the expression. What is an outermost operator? When there is only one operator within a selection box, it is the outermost operator. If the selection box contained the expression 5+2, pressing the (DEL) key would delete the plus sign (+). You can then enter any other legal mathematical operator, such as the minus (–), multiplication (*), or division (÷) sign.

When there is more than one operator within a selection box, the outermost operator is the operator that applies to all the numbers and variables in the selection box. For example, the minus sign is the outermost operator in the expression 3–4*8; the multiplication sign applies only to the 4 and 8. The multiplication sign is the outermost operator in the expression (2+3)*(5–4); the plus sign applies only to the 2 and 3, and the minus sign applies only to the 5 and 4.

When you want to replace a specific operator, simply click on that operator, press (DEL), and enter the new operator. For example, if you wanted to replace the addition operator in the expression (2+3)*(5–4) with a subtraction operator, you would click on the plus sign (+), press (DEL), and enter a minus sign (-).

It may not always be obvious which operator will be deleted when a selection box encloses a complex expression and you press the (DEL) key. If you don't like what happens when you press (DEL), select **Undo Last Edit** under the **Edit** menu.

In Example 2-4, you enter some text and an equation; in Example 2-5, you edit both the text and the equation.

EXAMPLE 2-4

Entering Text and an Equation

Enter the following text: `This is more example text.` Then evaluate the following expression, and create a function by setting the expression equal to $y(x)$:

$$\frac{5^2}{\sqrt{3^2+4^2}}+1$$

SOLUTION

If your document window is not already clear, clear it by selecting **Regions** and then **Select All** under the **Edit** menu (or by dragging the mouse across all regions) and then pressing (DEL).

Type: `"This is more`(ENTER)`example text.`

Remember, you must first type a double quote (") to let Mathcad know you are creating a text region and not entering a variable name in an equation region.

Click below the text region, and then type: 5^2

You want everything you have entered so far to be in a numerator (see the preceding expression), but if you simply press the division character (/) at this point, only the 2 will be in the numerator. To avoid this,

press: (↑)

A selection box, our third editing tool, appears around the 2. The next operation will apply to the 2, not what you want. Therefore

press: (↑) again.

The selection box now selects the 5^2. You are now ready to divide (/).

Type: /

A shortcut key that will create a radical (square root) is the backslash key (\).

Type: \3^2(↑)(↑)+4^2

Now you want to add 1 to the entire expression entered so far. You must create and expand a selection box that contains the entire expression and then add 1.

Type: (↑)(↑)(↑)(↑)(↑)

Notice the selection does contain the entire expression. You can press the (↓) key to decrease the size of the selection box. Do so now just to see the result and then use the (↑) key to put the box back where you want it. To finish,

type: +1=(ENTER)

You should now see the following on your screen:

This is more
example text.

$$\frac{5^2}{\sqrt{3^2+4^2}}+1=6$$

Click on the equals sign and press (DEL), (↓)

Press: (↑) until the entire expression is enclosed in a selection box (or press the spacebar once).

Press: (INS). Notice that the cut-off corner of the selection box moved to the left and that the selection box turned red.

Type: `:y(x)`(ENTER)

You should now see the following on your screen:

This is more
example text.

$$y(x) := \frac{5^2}{\sqrt{3^2+4^2}} + 1$$

The ability to edit equations easily is of critical importance when using Mathcad. For beginners, the challenge is to figure out how to make desired changes and then, with practice, to memorize what they have learned. For some more practice, work through Example 2-5, where you edit the regions created in the last Try It! exercise to create an entirely new document.

EXAMPLE 2-5

Editing Text and an Equation

In the last example, you entered

This is more
example text.

$$\frac{5^2}{\sqrt{3^2+4^2}} + 1 = 6$$

Edit the text to read, "This is excellent example text." Then edit the expression to evaluate

$$5^6 \cdot \sqrt{3^2 - 1.5} + 1$$

SOLUTION

Click on the word *more* within the text region.

A red insertion point should appear. You can move the insertion point anywhere within the text region using the arrow keys.

Use (BKSP) or (DEL) as appropriate to delete the word *more*.

Type: `excellent`

The text region grows as you enter new text.

Click the mouse on the division operator on the left side of the equation.

A selection box should appear containing the entire fraction to the left of the +1.

Press: (DEL) to delete the outermost operator, in this case, the division operator.

The division operator is replaced by a placeholder.

Type: *(ENTER)

You should now see

This is excellent
example text.

$$5^2 \cdot \sqrt{3^2 + 4^2} + 1 = 126$$

Click on the plus sign (+) under the radical, delete it, and replace it with a minus sign (-).

Click on the 4^2 under the radical, and edit it to read 1.5.

Click on the 2 in exponent of the 5, and edit it to read 6.

Press: (ENTER)

The screen should now show

This is excellent
example text.

$$5^6 \cdot \sqrt{3^2 - 1.5} + 1 = 4.279 \cdot 10^4$$

Notice that Mathcad automatically switched to scientific notation when the result of the calculation became relatively large.

Try It Clear the document window, enter the following text, and evaluate the following expression:

This is some
example text.

$$\frac{5 - 3 \cdot 6^2}{-4 + \sqrt{3^2 + 4^2}} + 1$$

Now edit the text to read, "This is an example of text." Then edit the expression to evaluate

$$\left[\left(5 + 3 \cdot 6^2\right) + \sqrt{3^3 + 4^4}\right] + 1.2$$

If you make a mistake, select **Undo Last Edit** under the **Edit** menu. As a last resort, you can simply clear the screen and start over.

Using Variables

Until now, you have been using Mathcad as a simple calculator. However, it is when Mathcad is performing algebra, calculus, and other more advanced mathematical operations that it really begins to shine. Moving forward from simple calculations starts with learning to use variables and functions.

In Mathcad, you enter variables into an equation or expression in the same way that you enter numbers. However, Mathcad will not be able to compute the value of an expression containing variables until you either tell it the values of the variables or give it a way to calculate the values itself. In this chapter, you will give Mathcad the values. You do this by pressing the colon (:) after typing the variable name. Mathcad adds an equals sign and then waits for you to type in a value or range of values for the variable.

Note that if you forget to use the colon and only press the equals sign, Mathcad will compute the value for you if it knows the values of all the variables used to the left of the equals sign. More likely, Mathcad will highlight the unknown variables, indicating it does not know their values and cannot compute the value of the expression.

An important note: Mathcad works left to right, top to bottom. Therefore, when a variable appears in an expression, you must define the value of that variable somewhere on the screen above or to the left of the expression. Example 2-6 demonstrates this principle and gives you some experience using variables.

EXAMPLE 2-6

Using Variables

Create two variables, *youngest* and *oldest*, and assign them the values 5 and 10, respectively. Find the average value of the two variables.

SOLUTION

First, clear the document window. A red crosshair should be present.

> Click the mouse halfway down the document window, along the left margin.
>
> Type: `oldest+youngest`(↑)(↑)`/2=`(ENTER)

You should see

$$\frac{youngest + oldest}{2} =$$

Note that Mathcad has highlighted the two variable names, indicating that it does not know their values and therefore could not compute the value of the equation when you pressed the equals sign.

> Click the mouse near the top of the document window, near the left margin.

You must define the values of *youngest* and *oldest* above or to the left of the expression that uses their values.

Type: `youngest:5`(ENTER)

Mathcad no longer highlights the variable name *youngest* in the expression because Mathcad now knows its value.

Type: `oldest:10`(ENTER)

Mathcad no longer highlights the variable name *oldest* in the equation and now computes the value of the expression. Your screen should now show the following:

$$youngest := 5$$

$$oldest := 10$$

$$\frac{youngest + oldest}{2} = 7.5$$

Try It Define a third variable called *middle* in Example 2-6 and set it equal to 11. Modify the expression in order to determine the average of the three variables. Note that *middle* must be defined above or to the left of the expression that computes the average. You can define it to the right of the definition of *youngest*.

Remember, you must define the values of all variables above or to the left of any expression or equation in which they appear. Were you to move the region containing youngest:=5 below or to the right of the expression, the expression value would disappear. Mathcad would no longer be able to compute the value. Also note that Mathcad makes a distinction between uppercase and lowercase letters in a variable name. The variable name *youngest* is not the same as the variable name *Youngest*; Mathcad sees them as two different variables.

Predefined Constants

Mathcad includes the following four built-in constants:

π	pi
e	the natural logarithm base
infinity	stored as 10^{307}
%	stored as 0.01

To enter π into an expression, press (CTRL)+p. To enter the natural logarithm base, simply enter e; Mathcad will interpret it as a constant, not a variable. To enter infinity, press (CTRL)+z. To calculate *p%* of a quantity, multiply the quantity by p and then multiply it by %. Example 2-7 demonstrates the use of all four constants.

EXAMPLE 2-7

Using Constants

Determine the area of a circle with a radius of 4.3. Then plot a graph of the area of a circle as a function of radius with radius ranging from 10 to 25 in increments of 0.2.

SOLUTION

First, clear the document window. Then:

Type: `radius:4.3`(ENTER)

Type: `area:`(CTRL)`+p*radius^2`(ENTER)

Remember, pressing (CTRL)+p displays the constant π.

Type: `area=`(ENTER)

You should see

radius := 4.3

area := $\pi \cdot$ radius^2

area = 58.088

Notice that the approach used here is different from the approach used in the previous example. You did not merely enter π*radius^2= but rather made it equal to the variable, *area*. You then asked Mathcad to compute the value of *area*. Now, to make the requested graph:

Type: `radius:10,10.2;25`(ENTER)

Define a function in which *radius* is the independent variable and *area* is the dependent variable.

Enter: `area(radius):`(CTRL)`+p*radius^2`(ENTER)

Select **Graphics** and then **Create X-Y Plot**.

In the blue selection box under the X-axis,

enter: `radius`

Click on the placeholder to the left of the Y-axis and in the resulting blue selection box,

enter: `area(radius)`(ENTER)

You should now see:

radius := 10, 10.2 .. 25

$\text{area}(\text{radius}) := \pi \cdot \text{radius}^2$

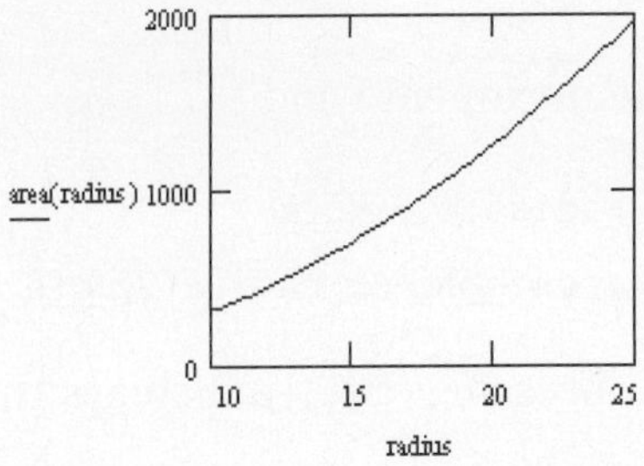

EXAMPLE 2-8

Determining a Percentage

A given capacitor effectively charges at the rate $v(t) = 14e^{-.2/t}$, where $v(t)$ is in volts, 14 is the fully charged value in volts, and t is in seconds. Determine the voltage on the capacitor when it is charged 20% of full charge.

SOLUTION

You can see from the function that as t increases, $-2/t$ approaches 0 and $e^{-2/t}$ approaches 1 ($e^0 = 1$). Therefore, as t increases, $v(t)$ approaches 14, meaning the full-charge voltage of the capacitor is 14 volts. You want 20% of that value.

Type: `20*%*14=`(ENTER)

You should see

$20 \cdot \% \cdot 14 = 2.8$

Try It An electronic signal dies out in a certain circuit at the rate: $v(t) = 0.4e^{-4t}\cos(t)$, where $v(t)$ is in volts and t is in seconds. Determine the time at which the signal has reached 10 percent of its value at $t = 0$.

Moving Regions

When you create regions within a document, they grow in size as you enter numbers and text. As a result, they may end up overlapping each other. Even if no regions overlap in your final product, you may not like the overall spacing of the regions. Therefore you might want to move them in order to produce a document that is more pleasing to the eye.

If you do not care about the appearance of the document and simply want to eliminate region overlapping, you can select **Separate Regions** under the **Edit** menu. However, if you want to move the regions, you can use **Cut** and **Paste** under the **Edit** menu, or you can use the mouse to drag the regions to new locations. Either way, you must first select the region to be moved.

When using **Cut** and **Paste**, you select the region by clicking within it and then pressing ↑ until the entire contents of the region are enclosed in a selection box. You can then **Cut** the region, click the mouse at the destination you desire, and perform the **Paste**. Pressing SHFT+DEL is a shortcut way to **Cut**, and pressing SHFT+INS is a shortcut way to **Paste**.

The mouse is used to select one region in the same way that it is used to select all regions; you simply drag the mouse across the desired region and then release the mouse button. The perimeter of the region appears as a dashed line. If the mouse cursor is moved within the box, a large black crosshair appears, indicating that the region now can be moved. Pressing and holding the mouse button within the region allows you to drag the region anywhere within the document. While the dashed line box is visible, you can also use **Cut** and **Paste** to move the region.

EXAMPLE 2-9

Moving a Region

Create an equation that computes the square root of 2, then move the region box that contains it.

SOLUTION

First, clear the document window. Then

> type: \2=ENTER
>
> Click the mouse on the equals sign (=) of the equation you just entered.

A selection box should appear around the entire equation. Clicking anywhere in the equation other than on the equals sign would have required that you press ↑ until a selection box enclosed the entire equation.

> Select **Cut** under the **Edit** menu.

The equation should disappear.

> Click any other place in the document window.
>
> Select **Paste** under the **Edit** menu.

The equation should reappear, but now with a dashed line around it, indicating the boundaries of the equation region.

> Click the mouse elsewhere in the document.

The red crosshair should return. The dashed line box should disappear.

> Press and hold down the mouse button near, but outside the region of, the equation. Drag the mouse cursor into the equation region and release the mouse button.

The dashed box should reappear.

Move the mouse cursor inside the equation region.

A large black crosshair appears.

Press and hold down the mouse button inside the dashed box and drag the equation wherever you like.

You can move all existing regions together, maintaining their relative positions, by selecting **Regions** and then **Select All** under the **Edit** menu and then using the mouse to drag the regions. You can move a group of adjacent regions by dragging the mouse across them, creating dashed boxes around them. You can then grab any one region by clicking within it, holding the mouse button down, and dragging it, along with the other boxed regions, to any new location.

Try It Enter the following equations and then reverse their position so that Mathcad can compute the area of the square:

$area := side^2$

$side := 2.3$

Note: Mathcad has highlighted *side* in the top equation because it needs to know its value in order to compute the value of *area*.

Error Messages

Mathcad may not be able to evaluate an expression because of an error in the expression. When this occurs, Mathcad will display an error message. A common cause of errors is not providing enough information. An example is when you press the (ENTER) key before providing values for all placeholders in an equation or before defining the independent and dependent variables for a graph. Other errors include dividing any number other than zero by zero (called *singularity* by Mathcad), incorrectly mixing arrays and scalars, and using incompatible units.

The message describes the surface problem, and you can always press (F1) to get more information. However, the real problem may not lie with the equation flagged but rather with an equation on which the flagged equation depended for information. For example, Mathcad does not attempt to evaluate the definition of a function until it encounters a use of that function in an equation. Therefore, if there is an error in the function definition, it is the equation that uses the function which will be flagged as having an error. Mathcad will not trace the source of the error to the function definition; you must find the error on your own. You will see how this works in Example 2-10.

EXAMPLE 2-10

Determining Error Messages

Determine the error message that Mathcad will generate if you divide a nonzero variable by 0 and if you divide zero by zero.

SOLUTION

First, clear the document window. Then

Type: `x:0`(ENTER)

Type: `y:1/x`(ENTER)

You should see

$$x := 0$$

$$y := \frac{1}{x}$$

singularity

Mathcad flagged the second equation as having involved a division by zero error, labeling it `singularity`. However, this equation is fine; the true problem is that x was given a value in a previous equation that is erroneously used in the second equation.

Type: `0/0=`(ENTER)

You should now see

$$x := 0$$

$$y := \frac{1}{x}$$

singularity

$$\frac{0}{0} = 0$$

No error was generated when you divided zero by zero.

Application 1

SIGNAL PROPAGATION

Electrical engineers involved in communications must often determine the time it takes a signal to propagate from its source to a destination. Among other parameters, this delay establishes the maximum rates at which two-way communication can take place, whether that communication be between two people or two computers. Let's apply the five-step problem-solving process to the problem of determining the approximate propagation delay between two points on the earth when a satellite is used as a relay station.

1. Define the Problem

Compute the approximate propagation delay of a radio signal traveling between two points on the earth when a satellite is used as a relay station.

2. Gather Information

The information required to solve this problem includes the speed at which a radio signal propagates in space and the distance at which a satellite orbits above the earth. The propagation speed in the atmosphere can be neglected because the atmosphere does not significantly affect the propagation speed of radio waves. The speed at which radio waves propagate in space is the speed of light, 186,000 miles per second. Geosynchronous communications satellites orbit at 22,300 miles above the earth.

3. Generate and Evaluate Potential Solutions

The distance between the two points on the earth between which communication is taking place can be neglected as long as they both have line-of-sight access to the satellite. This distance, which forms the base of a triangle with the satellite at the top vertex, has little effect on the 44,600 miles the signal must travel to the satellite and back to earth. Therefore, the propagation delay can be computed as the total distance traveled, 44,600 miles, divided by the speed, 186,000 miles per second.

4. Refine and Implement a Solution

Implementing a solution for this problem simply involves calculating 44,600 divided by 186,000. Using Mathcad to solve the problem,

$$\text{speed} := 186000$$

$$\text{distance} := 44600$$

$$\text{delay} := \frac{\text{distance}}{\text{speed}}$$

$$\text{delay} = 0.24$$

The result is measured in seconds.

5. Verify and Test the Solution

You have found that the propagation delay of a signal traveling between two points on earth via satellite is approximately 0.24 second. Testing this solution would involve using an accurate clock to measure the time it takes to receive a response to a signal sent to the destination. One would expect to record 0.48 second, the time it takes the signal to propagate to its destination plus the time it takes the response to propagate back.

What If

What if a signal had to travel through a coaxial cable from Los Angeles to New York and from there via satellite to Paris? Assume the distance between Los Angeles and New York is 3,000 miles and that signals travel at 80 percent of the speed of light over coaxial cables.

SUMMARY

This chapter covered the basics of using Mathcad. We began by describing the structure of Mathcad documents, explaining that they consist of a collection of equation, text, and graph regions. We then described how to use the pull-down menus to save and print documents. We introduced the subject of document creation by having you enter an equation, some text, and a graph. In some detail, the final section of this chapter covered the editing of documents. We described the use of the crosshair, insertion point, and selection box editing tools. We also included an explanation of how to use variables in your equations and how to use Mathcad's predefined constants such as π and e. The chapter ended with a brief discussion of error messages.

Key Words

assignment operator (:)
crosshair (red and black)
Electronic Book
evaluation operator (=)
font bar
insertion point
menu bar
outermost operator
palette
placeholder
QuickSheet
region (equation, graph and text)
selection box
shortcut key sequence
singularity
text band
title bar
toolbar

Problems

1. Select **QuickSheets...** under the **Help** menu. Select the **Graphing** Quicksheets. What function is plotted in the "Quick X-Y Plot of a Function" Quicksheet?
2. Select **Open Book...** under the **Books** menu, double click on the Deskref.hbk icon, and double click on <u>Table of Contents</u>. Scroll down the table of contents and double click on **1.2 Physical Constants.** What is the speed of light in a vacuum, as indicated here?
3. Determine the value of (3.2–1.7)/(5.2*8.2).
4. Determine the value of the square root of 5.7.
5. Determine the surface area and volume of a cube with a side length of 5.5.
6. Determine the side length and surface area of a cube with a volume of 10.
7. Determine the circumference and area of a circle with a radius of 10.
8. Determine the radius of a circle with an area of 6.7.

9. Determine the volume of a sphere with a radius of 9.2.
10. Determine the radius of a sphere with a volume of 13.4.
11. Determine the radius of a sphere with a surface area of 13.4.
12. Determine the area of a rhombus whose sides are 4 inches long and whose acute angle is 30°.

In each of Problems 13 through 15, enter three equations: the first to define the value of *X*, the second to define the value of *Y* in terms of *X*, and the third to compute *Y*.

13. Determine the value of Y when $Y = 2X$ and $X = 3.4$.
14. Determine the value of Y when $Y = X^2+X+3$ and $X = 4$.
15. Determine the value of Y when $Y = X^{3.2} - 2.4X^{-1.2} + 7.3$ and $X = 3.4$.
16. Determine the value of Y when $Y = 10e^{-5t}$ and $t = 3.2$.
17. Determine the value of Y when $Y = 1/(10e^{-5t})^{-1/3}$ and $t = 2.3$.
18. Given the quadratic equation

    $$Y = AX^2+BX+C$$

    in which $A = 3$, $B = 6$, and $C = 2$:

    a. Compute the values of the two roots of the equation. To do this, you must first define values for *A*, *B*, and *C*, and then define equations for the two roots, *X1* and *X2*, in terms of *A*, *B*, and *C*. Lastly, have Mathcad compute the values for *X1* and *X2*.

    b. Change the value of *A* to 0 in the region that defines its value. Press (ENTER) to effect the change. What occurs?

    c. Move the region that defines the value of *B* to the right of the equation that defines *X2*. What occurs?

    d. In the region that sets the value of *C*, change the variable name from *C* to a lowercase *c*. Press (ENTER) to effect the change. What occurs?

Use the five-step problem-solving process while working on the remaining problems.

19. Determine the propagation delay of a signal that must travel around the earth at the equator in a coaxial cable. Assume that the propagation speed of a signal in a coaxial cable is 80 percent of the speed of light.
20. Determine how long it will take an object with no air resistance to hit the ground falling from a height of 1,000 meters.
21. Assume a bullet with no air resistance is shot straight up, exiting the muzzle of the gun at a velocity of 5,000 feet per second. How long will it take the bullet to hit the ground?
22. Assume a bullet with no air resistance is shot horizontally at a height of 4 feet, exiting the muzzle of the gun at a velocity of 5,000 feet per second. How long will it take the bullet to hit the ground?

3 Engineering and Scientific Computations

Information Superhighway

Its official name is the National Information Infrastructure (NII), but it is more often called the "information superhighway." Although there is some argument over what it is, or will be, most agree that its main thoroughfare is the Internet, a worldwide network of thousands of interconnected computer networks. These computer networks, each typically including hundreds of PCs and terminals connected to mainframes, are owned by universities, government agencies, the military, and commercial companies. Individual users access the Internet using either a computer terminal or a PC to access a mainframe that is linked into the Internet by cable, microwave, or satellite.

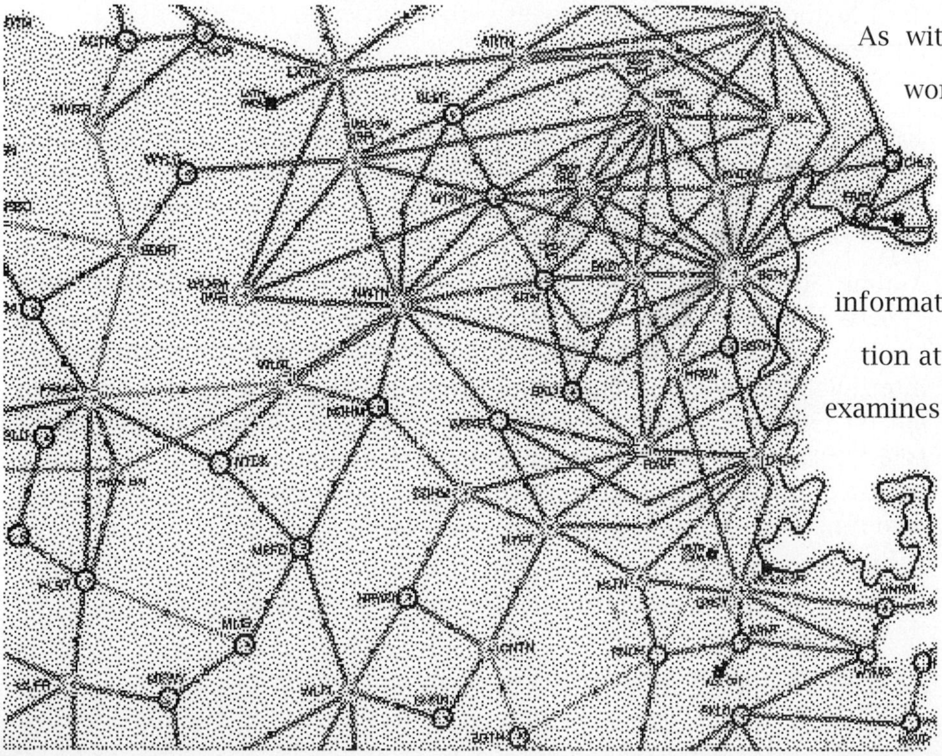

As with most computer networks, the two main goals of the Internet are to provide communication capability and information access. An application at the end of this chapter examines one aspect of computer network design.

INTRODUCTION

In Chapter 2, you learned the basics of using Mathcad, including how to get the program running and how to save and print files. Chapter 2 also described Mathcad's editing features and showed you how to use variables and predefined constants in your equations. In this chapter, we will discuss many of Mathcad's more advanced features. The chapter starts by showing you how to use Mathcad's built-in functions, including its trigonometric, logarithmic, and exponential functions, in your equations. It then discusses how to create and use your own mathematical functions. A description of how to use units, such as meters and seconds, in your equations follows. Here the chapter briefly explains how Mathcad tracks your units, ensuring that you use them correctly, and provides the correct units for the results of your calculations. The creation of tables of data is the next subject. Here, you are shown how to perform iterative tasks.

The second section of the chapter describes how to create, edit, and compute with one-dimensional arrays (vectors) and two-dimensional arrays (matrices). The section includes a description of available array operators and functions, as well as a discussion of how to read and write arrays in data files.

The final section of the chapter explains how to use most of Mathcad's more popular features. These features include the solving of single and sets of equations, computing summations and products, generating random numbers, sorting vectors and matrices, creating histograms, and computing statistical functions. The chapter concludes with an application problem that applies the five-step problem-solving process introduced in Chapter 1, using some of the Mathcad features described in this chapter.

It is important to remember that Mathcad's capabilities go well beyond that which is presented here. Mathcad's list of built-in functions is extensive. And if these functions are not enough, with Mathcad 6.0 Plus you can create your own functions and write complete programs using Plus's powerful built-in programming capabilities. (Mathcad 6.0 Plus includes many advanced built-in functions and some capabilities not available in standard Mathcad 6.0.) See your user's manual or the Mathcad help features for more information.

3-1 MATHEMATICAL FUNCTIONS

Using Built-In Functions

Among Mathcad's more than 100 built-in functions are trigonometric, hyperbolic, logarithmic, exponential, and Bessel functions. Also available are Fourier transforms and functions for performing statistical analysis, interpolation, linear regression, and sorting. Other built-in functions operate on complex numbers, vectors, matrices, and files. An easy way to select a function is click on the Choose Function (showing an *f(x)* icon) button on the toolbar. Mathcad will then alphabetically list all of the available functions, along with descriptions of their use.

Extensive examples of the use of Mathcad's many functions can be found in two places. The first is found by selecting **QuickSheets...** under the **Help** menu. If you then select **Arithmetic**, for example, you can find example problems using trigonometric functions. To access the second source of help, select the **Books** menu and then select **Open Book....** Double click on the Deskref.hbk icon and then double click on Table of Contents. Again, help is offered on many subjects, from using physical constants to solving calculus problems.

The following is a list of Mathcad's most commonly used trigonometric, logarithmic, and exponential functions.

Trigonometric functions (require arguments in radians; inverse functions return values in radians)

Function	Inverse
sin(z)	asin(z)
cos(z)	acos(z)
tan(z)	atan(z)
csc(z)	
cot(z)	
sec(z)	

Logarithmic and exponential functions

ln(z)—Returns the natural logarithm of z
log(z)—Returns the base 10 logarithm of z
exp(z) or e^z—Returns e raised to the power z

All of Mathcad's built-in functions are straightforward and simple to use. You simply type the common name for the function. For example, if you want the square of the *cosine* of 2*t*, you type

```
cos(2*t)^2=
```

Note, that $\cos(2t)^2$ equals $[\cos(2t)]^2$ and not $\cos[(2t)^2]$.

Example 3-1 demonstrates the use of two built-in functions, *sin* and *cos,* and two predefined constants, e and π.

EXAMPLE 3-1

Using Built-In Functions

Evaluate the function $f(t) = e^{-t}(\cos 2\pi t + \sin 2\pi t)$ for $t = 0.25$.

SOLUTION

If your document window is not already cleared, clear it now. The red crosshair should appear. It should be in the upper left-hand corner. If not, click the mouse anywhere in the window where you want to begin; the red crosshair should appear. If it does not, select **New** under the **File** menu to open a new document.

Give *t* its specified value using the assignment operator, : (the colon).

Type: `t:0.25`(ENTER)

Note that, on the screen, Mathcad inserted an equals sign after the colon. Together, the colon and equals sign indicate that you, and not Mathcad, provided the information to the right.

Using CTRL+p to enter the value π, define the function.

Type: `e^-t↑↑↑*(cos(2*`CTRL`+p*t)+sin(2*`CTRL`+p*t))=`ENTER

Typing ↑ three times was necessary so that Mathcad applied the remainder of the expression to e^{-t} and not to just *t*.

Since you followed the expression with the evaluation operator, =, Mathcad provided the value of the expression. You should see

$$t := 0.25$$

$$e^{-t}\cdot(\cos(2\cdot\pi\cdot t) + \sin(2\cdot\pi\cdot t)) = 0.779$$

Try It Evaluate *mg*(*tan*(*a*)) for $m = 5$, $g = 9.8$, and $a = \pi/8$. If *m* is in meters, *g* is in m/sec^2, and *a* is in radians, what are the units of the result? (Later in this section, you will learn how to have Mathcad determine this for you.)

Creating Your Own Functions

You can define your own functions in Mathcad by entering a function name, one or more independent variables within parentheses, a colon (the assignment operator), and the mathematical expression that uses the independent variable(s) to compute the function result. For example, to define the function $f(x)$ as x^2, you would type

`f(x):x^2`ENTER

In Example 3-2, you learn how to create a user-defined function that computes the altitude of a free-falling body that begins its fall 1,000 meters above the earth.

EXAMPLE 3-2

Using Your Own Functions

Determine the altitude after 10 seconds of a free-falling object that began its fall at rest at an altitude of 1,000 meters above the ground. Assume the body has no air resistance.

SOLUTION

Although the value of *acc*, the acceleration due to the force of gravity, could have been entered directly into the definition of the function *position*(*t*), this solution begins by defining *acc* as a constant to show you an alternative approach.

Type: `acc:9.8`ENTER

Now you will define your own function, called *position*(*t*), that computes the altitude of the falling object at time *t*. Again, you use the assignment operator, :

Type: `position(t):1000-acc/2↑↑*t^2`ENTER

You should now see

$acc := 9.8$

$position(t) := 1000 - \frac{acc}{2} \cdot t^2$

You can now compute the altitude of the body for any value of *t*. To obtain its position at $t = 10$ seconds, as specified in the problem,

type: `position(10)=`(ENTER)

Your document window should now show

$acc := 9.8$

$position(t) := 1000 - \frac{acc}{2} \cdot t^2$

$position(10) = 510$

The altitude of the object after 10 seconds of free fall is 510 meters above the ground; it has not even fallen halfway yet. (By clicking on the 10, you can edit the value of *t* to find the altitude at other times. Pressing (ENTER) will instruct Mathcad to compute the new result.)

Save your work in example.mcd; it will be used in the Example 3-4.

Try It

Create a function *position*(*t*) that computes the altitude at time *t* in seconds of a free-falling object that began its fall with a velocity of 100 meters/second at an altitude of 1,000 meters above the ground. Assume the body has no air resistance.

Use *position*(*t*) to determine the altitude of the object after 5 seconds of free fall.

Graphing a Function

The creation of graphs is covered in detail in Chapter 4, where the topics include the graphing of multiple functions and vectors as well as the creation of histograms and polar, surface, contour, and scatter plots. The following example offers practice in the graphing of single functions.

EXAMPLE 3-3

Graphing a Function

Graph the function $f(t) = e^{-t}(\cos 2\pi t + \sin 2\pi t)$.

SOLUTION

In Example 3-1, you gave the variable *t* a specific value. To graph *f* as a function of *t*, you must give *t* a range of values. For example, if you wanted to plot the graph of *f* for *t* equals 0 to 5,

type: `t:0;5`(ENTER)

Then define the function *f*(*t*):

Type: `f(t):e^-t`(↑)(↑)(↑)`*(cos(2*`(CTRL)`p*t)+sin(2*`(CTRL)`p*t))`(ENTER)

Then create the graph:

Select **Create X-Y Plot** under the **Graphics** menu (or use the Graphing Palette button on the palette).

An X-Y graph should appear with placeholders along the bottom (the X-axis) and left-hand side (the Y-axis).

A selection box (discussed in Section 2-4) appears around the placeholder below the center of the X-axis because Mathcad expects you to enter the independent variable first. Now:

Designate the independent variable in the selection box by typing: `t`

Click on the placeholder to the left of the center of the Y-axis.

Designate the dependent variable by typing: `f(t)`(ENTER)

You should now see

$t := 0..5$

$f(t) := e^{-t} \cdot (\cos(2 \cdot \pi \cdot t) + \sin(2 \cdot \pi \cdot t))$

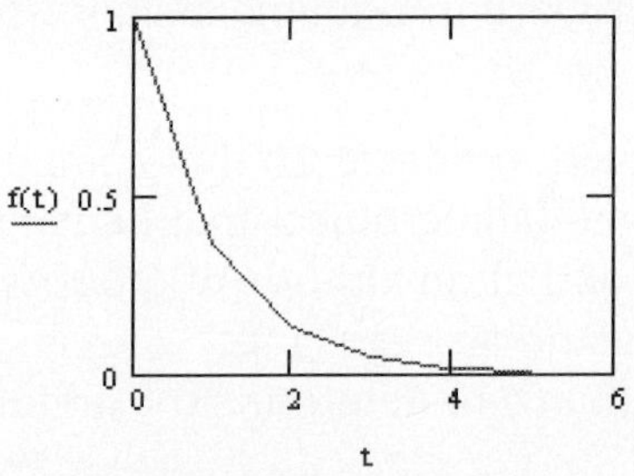

For more information on graphing, select the **Help** menu and then select **QuickSheets....** Help is offered on graphing as well as many other subjects.

Try It Graph the function *f(a)* = *mg*(*tan*(*a*)) for *m* = 5, *g* = 9.8, and *a* = 0 to 6.

Adding Units

As noted earlier, Mathcad allows you to use units with your constants and variables. It will check for consistency in your use of units and compute the appropriate units for the results of calculations that use units. To view a list of the available units, click on the Units button (showing a measuring cup icon) on the toolbar.

Mathcad provides three built-in systems of units:

MKS units—the base units are the meter, kilogram, and second.

CGS units—the base units are the centimeter, gram, and second.

British units—the base units are the foot, pound, and second.

Mathcad uses the MKS (SI is the preferred term used today) system of units by default; to choose one of the other two systems of units, select **Units** under the **Math** menu.

In addition to the base units, under the MKS system, Mathcad provides the user with approximately 80 derived units (see the appropriate User's Guide Appendix). A partial list of these derived units includes centimeters (cm), grams (gm), minutes (min), hours (hr), days (day), years (yr), coulombs (coul), radians (rad), degrees (deg), hectares (hectare), liters (liter), kilometers/hour (kph), newton (newton), joule (joule), calorie (cal), Pascal (Pa), volt (volt), amp (amp), watt (watt), ohm (ohm), and Hertz (Hz). You can even use inches (in), feet (ft), miles (mi), acres (acre), miles/hour (mph), horsepower (hp), pounds/square inch (psi), and other British units while using the MKS system of units.

In order to use units within any of your calculations, simply multiply a quantity by the units you desire. For example, you can define the acceleration due to gravity by typing

```
accel:9.8*m/sec^2
```

In Example 3-4, you add units to your work of the previous example.

EXAMPLE 3-4

Using Units

Retrieve your Example 3-2 document by opening example.mcd. Add appropriate units to your Example 3-2 work. That is, determine the altitude after 10 seconds of a free-falling object that began its fall at rest at an altitude of 1,000 meters above the ground. Assume the body has no air resistance.

SOLUTION

Start by deleting or adding work in your document window until it includes only the following:

$$acc := 9.8$$

$$\text{position}(t) := 1000 - \frac{acc}{2} \cdot t^2$$

Click at the right end of the first equation, the one that defines acceleration (acc).

An insertion point should appear as a blue vertical line.

Multiply the 9.8 by the appropriate units by typing: `*m/sec^2`

If the two equation regions overlap, some part of the top equation might be obscured. If this is the case, under the **Edit** menu, select **Regions** and then **Separate**. The screen should now show

$$acc := 9.8 \cdot \frac{m}{sec^2}$$

$$\text{position}(t) := 1000 - \frac{acc}{2} \cdot t^2$$

The only remaining constant that has not had its units defined is 1,000 in the second equation, the one that defines the function *position*(*t*).

Give the 1,000 units in meters by multiplying it by m.

Click to the right of 1,000 and type: `*m`

Variables must also have units. Therefore, you must multiply t in the function definition by sec (seconds). You do not give units to the independent variable t in the function name *position*(*t*). Note that, in the function definition, you must enclose both sec and t in parentheses so that the product of the two is squared: $(t^*\text{sec})^2$.

Adding parentheses as appropriate, multiply t by sec.

The screen should now show

$$acc := 9.8 \cdot \frac{m}{sec^2}$$

$$position(t) := 1000 \cdot m - \frac{acc}{2} \cdot (t \cdot sec)^2$$

Let us again compute the position after 10 seconds.

Click below the second equation, leaving enough room for a third equation.

Without pressing (ENTER), type: `position(10)=`

The screen should show

$$acc := 9.8 \cdot \frac{m}{sec^2}$$

$$position(t) := 1000 \cdot m - \frac{acc}{2} \cdot (t \cdot sec)^2$$

$$position(10) = 510 \cdot m$$

Note that Mathcad automatically gave the units in meters. You can change the units of a result; simply click on the result and enter the new units in the placeholder that appears. For example:

Click on the result, 510 m.

Click on the placeholder that appears to the right of 510 m.

Type: `ft`(ENTER).

The screen should now show

$$acc := 9.8 \cdot \frac{m}{sec^2}$$

$$position(t) := 1000 \cdot m - \frac{acc}{2} \cdot (t \cdot sec)^2$$

$$position(10) = 1.673 \cdot 10^3 \cdot ft$$

The two results, 1,673 feet and 510 meters, are equivalent.

For more information on using units, access the Electronic Book on that subject by selecting **Open Book...** under the **Books** menu and then browsing through the table of contents.

Try It

Determine the altitude of a projectile at $t = 10$ seconds in meters and in feet given that its altitude is defined by

$$\text{position}(t) := 490 \cdot m - \frac{9.8 \cdot \frac{m}{sec^2}}{2} \cdot ((t - 10) \cdot sec)^2$$

Creating Tables

The work of engineers often involves the generation and analysis of tables of data. In order to create a table, you must first create a *range variable* by declaring a variable and assigning it a starting and ending value, separated by a semicolon. For example, to define a range variable *i* that varies from 0 to 10, type

```
i:0;10(ENTER)
```

An increment of 1 is assumed unless you specify otherwise. For example, to create a range variable that varies from 1.0 to 5.0 in increments of 0.1, type:

```
i:1.0,1.1;5(ENTER)
```

The 1.2 that follows the comma tells Mathcad that the second value of *i* in the series is 1.2, implying an increment of 0.1.

Example 3-5 demonstrates how to use a range variable to create a table. In this example, you will add a table to the work of the previous example, listing the altitudes of an object falling under the force due to gravity.

EXAMPLE 3-5

Using Tables

Create a table that displays the time and position of a free-falling object for $t = 0$ to $t = 1$ second in 0.1 second increments. Assume that at $t = 0$ seconds, the body is at rest at an altitude of 1,000 meters. Assume the body has no air resistance.

SOLUTION

If you saved your previous work in example.mcd, retrieve that file.

Define *acc* and *position*(*t*) as follows:

$$acc := 9.8 \cdot \frac{m}{sec^2}$$

$$\text{position}(t) := 1000 \cdot m - \frac{acc}{2} \cdot t^2$$

In order to create a table of values of *position*(*t*), you must first define a range of values for *t*. To do this for $t = 0$ to $t = 1$ in 0.1 second increments,

type: `t:0*sec,0.1*sec;1*sec(ENTER)`

The semicolon (;) becomes .. on the screen to indicate the definition is of a range variable. Only the third *sec* is necessary, so you could have entered `t:0,0.1;1*sec`(ENTER).

You should now see

$$acc := 9.8 \cdot \frac{m}{sec^2}$$

$$position(t) := 1000 \cdot m - \frac{acc}{2} \cdot t^2$$

$$t := 0 \cdot sec, 0.1 \cdot sec .. 1 \cdot sec$$

Now, to create the required table,

without typing (ENTER), type: `t=`

The equals sign does not appear, but a table of values of *t* does.

Click to the right of the t you just typed, and type:
`position(t)=`

Again, when you pressed the equals sign, the table appeared, instead of the evaluation operator, =, and the screen should now show

$$acc := 9.8 \cdot \frac{m}{sec^2}$$

$$position(t) := 1000 \cdot m - \frac{acc}{2} \cdot t^2$$

$$t := 0 \cdot sec, 0.1 \cdot sec .. 1 \cdot sec$$

t	position(t)
0·sec	$1 \cdot 10^3 \cdot m$
0.1·sec	999.951 m
0.2·sec	999.804 m
0.3·sec	999.559 m
0.4·sec	999.216 m
0.5·sec	998.775 m
0.6·sec	998.236 m
0.7·sec	997.599 m
0.8·sec	996.864 m
0.9·sec	996.031 m
1·sec	995.1 m

Together, the two tables display the times and altitudes of the object corresponding to the times $t = 0$ to $t = 1$ second in 0.1 second increments. You may want to save this work in example.mcd; it will be used in the next Try It!

You cannot convert the units in a table in the same way that you can with a scalar answer. For example, dividing the entry *position*(*t*) above the right table by ft (feet) would result in the table listing its distances in feet; however, the term m disappears from the table entries, and ft does not replace it.

To format the values in a table, double-click on the table. A dialog box will appear that gives you several formatting options. For example, you can display values in decimal, octal, or hexadecimal form, and you can set the precision of the values.

Mathcad shows only the first 50 values of a range of numbers in a table. If you require a list longer than 50 values, you must use more than one range variable. For example, in Example 3-5, you might enter `t1:0,0.01;0.49` and `t2:0.50,0.51;0.99` and then create two function definitions, *position*(*t*1) and *position*(*t*2), and two tables by typing the expressions *position*(*t*1)= and *position*(*t*2)=.

If you want to delete a table, click on the table and press the DEL key. When you click on a table, a selection box will appear, enclosing both the table and the expression that defined the table. Remember that pressing DEL whenever a selection box is present deletes the outermost operator within the selection box. In this case, the outermost operator is the (invisible) equals sign; deleting it removes the table. If you want to delete both the table and the expression that defined it, you should press the SHFT+DEL keys when the selection box encloses both the table and the defining equation. This action always deletes the entire contents of any selection box. Or, as noted earlier, you can drag the mouse across the table, defining the region with a dashed box, and press DEL.

Try It Retrieve the work of the previous example from example.mcd if necessary. Repeat the example, using U.S. customary units. That is, change the acceleration to 32 ft/sec^2 and convert the starting position to feet. The position table should list positions in feet.

Format the position table so that the displayed precision is 0, that is, the values in the table are rounded to whole numbers.

3-2 VECTORS AND MATRICES

A single number is referred to as a *scalar*, a one-dimensional array of numbers as a *vector*, and a two-dimensional array of numbers as a *matrix*.

Vectors

There are several ways to create a vector. The easiest way to create small vectors is to define an index using a range variable and then create the vector by typing its name, typing the left bracket character (`[`), and then entering its element values separated by commas. For example, to create a 6-element vector of zeros:

Type: `i:0;5`ENTER

Type: `X[i:0,0,0,0,0,0`ENTER

Array indices begin at 0. However, you can change the starting array index by specifying the origin value in the **Built-in Variables** selection under the **Math** menu. Many engineers prefer to have the starting index set at 1.

Using the left bracket again, you specify a particular vector element by typing *X[i* where **X** is the name of the array and *i* is a subscript, that is, the vector index. The index can be a variable. For example, Mathcad will report the value of the 6th element (X_0 being the first element) of the vector **X** if you enter *X[5=*.

A more general way to create a vector is to first select **Matrices** under the **Math** menu. A dialog box will appear, as shown in Figure 3-1.

Figure 3-1
Matrices Dialog Box

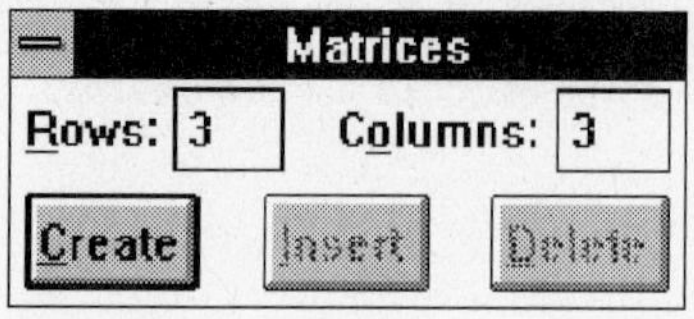

The default matrix size is 3 × 3; however, you can enter any size you desire in the Rows and Columns boxes. By default, Mathcad works with column vectors although you can create row vectors. If you want to create a column vector, enter the number of elements in the Rows box and enter 1 in the Columns box. Do the reverse to create a row vector.

If you are going to need several vectors, you can leave the dialog box open. You would click on the Create button in the dialog box each time you want to create a new array, changing the number of rows or columns if necessary. The dialog box may show the Create button as being inactive, but if you click on it, it will become active. Then click on it again to create an array of the size indicated by the Rows and Columns boxes.

Either before or after you have specified the size of a vector, click at the location where you want the array to appear in the document window. When you click on the Create button, an array is then created at the location of the red crosshair. It will be filled with the appropriate number of placeholders. For example, to create a 4-element column vector, after setting up the Matrices dialog box for a 4-row, 1-column matrix, do the following:

> Without pressing (ENTER), type: `X:`
>
> Click on **Create** in the **Matrices** dialog box.

Press (ENTER) and you should see

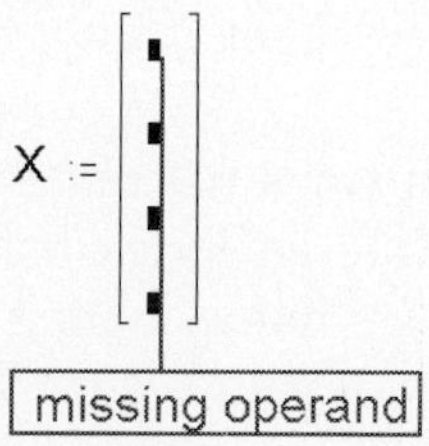

Pressing (ENTER) caused the error message, `missing operand` because Mathcad is waiting for you to enter the vector values. The fastest way to do this is to click on the highest placeholder (in a column vector), placing a selection box around it. Enter a value for that array element, and then press (TAB). Each time (TAB) is pressed, the selection box moves to the next placeholder, working its way down the column.

You can easily change the size of a vector by clicking on the vector so that it is enclosed in a selection box. Then, in the Matrices dialog box, set the number of rows or columns you want to add or delete. Lastly, click on either the Insert or Delete button in the Matrices dialog box. In this way, a vector can be changed into a matrix.

Using the Matrices dialog box, you cannot create arrays (vectors or matrices) with more than 100 elements. However, you can create arrays with up to 8 million elements, if you have enough memory in your computer, by reading the elements from a disk file. We will discuss reading and writing data files in Section 3-3.

A vector in which the elements are mathematically related can be created using a range variable. This saves you the effort of entering the vector elements one at a time. It also allows you to easily assign units to the vector elements, as you see in Example 3-6.

EXAMPLE 3-6

Creating a Vector

Create a 10-element vector that contains the even numbers, 2, 4, 6, ...20, with the units in meters.

SOLUTION

First define the range variable. It should start at 0 because Mathcad assumes the index of the first element in a vector to be 0.

Type: `i:0;9`(ENTER)

Now define the vector contents. The algorithm for establishing the elements is something you must work out yourself. In this case, the algorithm must produce the even numbers from 2 to 20. The *m is added to make the units meters.

Type: `X[i:(1+i)*2*m`(ENTER)

To test our algorithm to ensure the contents of **X** are what is expected,

type: `X[i=`

The equals sign is not shown, and you should now see

$i := 0 .. 9$

$X_i := (1 + i) \cdot 2 \cdot m$

X_i

2·m
4·m
6·m
8·m
10·m
12·m
14·m
16·m
18·m
20·m

See the QuickSheets for more information on vectors.

Graphing a Vector

Plotting the contents of a vector is similar to plotting a function. You define a range variable that is the index of the vector, create the vector if it does not already exist, and then create the graph. You must use the index as the independent variable of the graph; the vector name is entered as the dependent variable.

EXAMPLE 3-7

Graphing a Vector

Plot the contents of the vector created in Example 3-6, that is, a 10-element vector that contains the even numbers, 2, 4, 6, ...20.

SOLUTION

Since **X** contains 10 elements, define an index that ranges from 0 to 20 in steps of 2. Remember, after the comma you enter the second value in the series, not the step size.

Type: `i:2,4;20`

Now create **X**, this time entering the elements manually, separating them with commas.

Type: `X[i:2,4,6,8,10,12,14,16,18,20`(ENTER)

Notice that a column vector was created and shown in a box. The [told Mathcad you were defining a vector and that *i* is its index.

Now create the graph:

Click to the right center of the X vector and select **Create X-Y Plot** under the **Graphics** menu.

In the placeholder at the bottom center of the graph, type: `i`

Click on the placeholder to the left of the center of the Y-axis.

Type: `X[i`(ENTER)

The screen should now show

$i := 2, 4 .. 20$

$X_i :=$

2
4
6
8
10
12
14
16
18
20

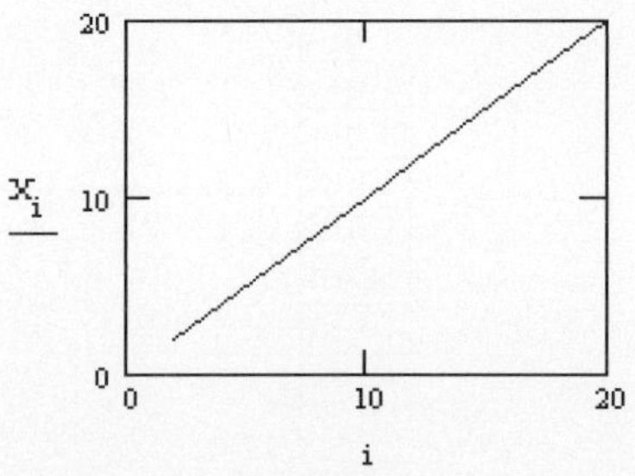

Try It Plot the contents of a 20-element vector whose elements are 1/*i* where *i* is the index, which has a range of 1 to 20.

Matrices

The easiest way to create matrices is to use the Matrices dialog box. Enter the size of the matrix you desire in the Rows and Columns boxes. If you are going to need several matrices, you can leave the dialog box open and reuse it.

As with vectors, either before or after you have specified the size of a matrix, click at the location where you want the matrix in the document window. When you click on the Create button, an array is created at the location of the red crosshair. It will be filled with the appropriate number of placeholders.

As with vectors, indices begin at zero; therefore, the element in the upper left-hand corner of a matrix would have the indices 0,0. Matrix elements are specified using two indices separated by a comma, for example, $X[i,j$.

EXAMPLE 3-8

Creating a Matrix

Create a 4 × 2 matrix **y**, in which the values of the elements are the same as the row number in which they reside. Give the elements the units feet.

SOLUTION

To create a matrix, you enter the name of the array, press the colon (:), and then click on the Create button in the **Matrices** dialog box.

Click in any clear area of the document window, preferably in the upper left-hand corner.

Without pressing ENTER, type: `y:`

Open the Matrices dialog box by clicking on Matrices under the Math menu.

Set the number of rows to 4 and the number of columns to 2 in the Matrices dialog box.

Click on the Create button in the Matrices dialog box.

Press: ENTER

You should now see

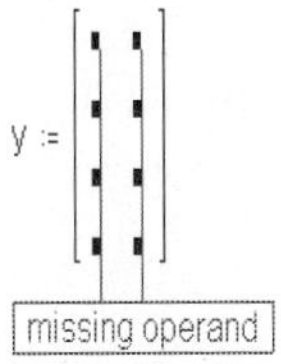

Click on the upper left placeholder in the matrix.

Enter the value `0,` since this element is in row 0, and press TAB.

The selection box should move down to element $y_{1,0}$ (row 1, column 0).

Enter the value `1,` since this element is in row `1,` and press TAB.

Continue until the array is filled with the specified values.

Place a selection box around the matrix and type: `*ft`ENTER

To place a selection box around the matrix, click anywhere within the matrix and press ⇧ until you have the selection box you desire.

To determine the element in **y** that is in row 2 and column 1:

Type: `y[2,1=`(ENTER)

You should see

$$y := \begin{bmatrix} 0 & 0 \\ 1 & 1 \\ 2 & 2 \\ 3 & 3 \end{bmatrix} \cdot ft$$

$$y_{2,1} = 0.61 \cdot m$$

The result is in meters (0.61 meter equals 2 feet) because meters are the default units used by Mathcad when dealing with distances.

See the QuickSheets for more information on matrices.

Try It

Create a 4 × 4 identity matrix called **I**. An identity matrix is a square matrix in which all elements are 0 except for the diagonal values from the upper left to the lower right, which are 1s. A square matrix is a matrix that has the same number of rows as it has columns. An identity matrix gets its name from the fact that, when it is multiplied by a matrix **X** that has the same number of columns as does the identity matrix, the result is **X**.

Array Computations

Mathcad can perform any standard array operation, including not only addition, subtraction, multiplication, and division of matrices, but also the transpose, determinant, inverse, dot product, and cross product operations. We will discuss all these operations in the following pages. Other operations include functions for determining everything from the minimum and maximum elements in an array to finding the eigenvalues of a square matrix and the Fast Fourier Transform (FFT) of a vector. Two of the simpler matrix operations, scalar multiplication and matrix addition, are examined in Example 3-9.

EXAMPLE 3-9

Using Array Mathematics

Assume the following matrix element represents 16 voltage samples taken from four voltage sensors in an electronic circuit:

$$X = \begin{bmatrix} 7 & 0 & 1 & -3 \\ 0 & 1 & 8 & 5 \\ -1 & 2 & 7 & 4 \\ -1 & 9 & 6 & 3 \end{bmatrix}$$

Row 0, the top row, stores samples from sensor 0, row 1 stores samples from sensor 1, and so on. Assume a circuit simulation in which these voltages are passed through a 4-channel amplifier. Sensor 0 feeds channel 0, sensor 1 feeds channel 1, and so on. Each channel has a voltage gain of 2. Channel 0 adds no voltage shift, but channel 1 adds a voltage shift of -1 volt, channel 2 adds +1 volt, and channel 3 adds +2 volts. That is, each channel doubles each input voltage and then adds an integer to it. Determine the sequence of outputs of the four channels. Lastly, determine the minimum and maximum output voltages from the 4-channel amplifier, given the 16 voltage samples.

SOLUTION

Clear the document window and place the red crosshair near the upper left-hand corner, but leave enough room for the matrix X.

Without pressing (ENTER), type: `X:`

Open the Matrices dialog box by selecting Matrices under the Math menu.

Set both the Rows and Columns in the Matrices dialog box to 4.

Create X by clicking on the Create button in the Matrices dialog box.

Enter the specified values into the elements of **X**.

To the right of **X**, create a 4×4 matrix **Y** that contains the voltage shifts:

$$\mathbf{Y} = \begin{bmatrix} 0 & 0 & 0 & 0 \\ -1 & -1 & -1 & -1 \\ 1 & 1 & 1 & 1 \\ 2 & 2 & 2 & 2 \end{bmatrix}$$

To add the gain and the voltage shift and display the output samples of the four channels:

Click below **X** and type: `X*2+Y=`(ENTER)

To determine the minimum and maximum voltage outputs,

below **X*2+Y**, type: `min(X*2+Y)=`(ENTER)

and then type: `max(X*2+Y)=`(ENTER)

If you label the four channel outputs as shown below, you should see

$$X := \begin{bmatrix} 7 & 0 & 1 & -3 \\ 0 & 1 & 8 & 5 \\ -1 & 2 & 7 & 4 \\ -2 & 9 & 6 & 3 \end{bmatrix} \qquad Y := \begin{bmatrix} 0 & 0 & 0 & 0 \\ -1 & -1 & -1 & -1 \\ 1 & 1 & 1 & 1 \\ 2 & 2 & 2 & 2 \end{bmatrix}$$

$$X \cdot 2 + Y = \begin{bmatrix} 14 & 0 & 2 & -6 \\ -1 & 1 & 15 & 9 \\ -1 & 5 & 15 & 9 \\ -2 & 20 & 14 & 8 \end{bmatrix} \begin{matrix} \text{<== ouput of channel 0} \\ \text{<== ouput of channel 1} \\ \text{<== ouput of channel 2} \\ \text{<== ouput of channel 3} \end{matrix}$$

$\min(X \cdot 2 + Y) = -6$

$\max(X \cdot 2 + Y) = 20$

Try It Create the following matrix:

$$\mathbf{X} = \begin{bmatrix} 1 & 2 & 7 & 0 & 1 & -3 \\ -3 & 1 & 0 & 1 & 8 & 1 \\ 2 & -2 & 1 & 2 & 1 & 4 \\ 0 & 2 & -2 & 1 & -2 & 3 \end{bmatrix}$$

- Multiply **X** by -3 and add 2.
- Determine the minimum element in **X**.
- Determine the maximum element in **X**.

Mathcad easily performs a variety of other operations on vectors and matrices. For example, to determine the transpose of a matrix **M**, type M(CTRL)1=(ENTER). To determine the determinant of a square matrix **M**, press the vertical bar key (|) before and after the matrix name: |M|=(ENTER). To determine the inverse of a matrix **M**, type: M^-1=(ENTER). To determine the dot product of two vectors **m** and **n**, type m*n=(ENTER). To determine the cross product of two 3-element vectors **M** and **N**, the only size array on which Mathcad will perform this operation, type M(CTRL)8N=(ENTER).

Reading and Writing Vectors in Data Files

Engineers must often collect and analyze large amounts of data. These data often are stored in computer *data files* by data acquisition systems and computer programs. Therefore, the ability to read and write data files is an important Mathcad feature. Mathcad reads data files into scalars, vectors, and matrices, and stores scalars, vectors, and matrices in data files.

The numeric data values in the files Mathcad reads must be ASCII characters that are separated by spaces, commas, or carriage returns. These data can include columns of numbers saved as a text file by a word processing program and files created by spreadsheets or computer programs. Mathcad saves data in files as ASCII characters, separated by spaces and carriage returns. Other applications, such as word processors and spreadsheets, can read Mathcad data files.

Mathcad includes two sets of file-handling functions, the names of which must be typed entirely in uppercase. *READ*, *WRITE*, and *APPEND* are used to read or write single data values, usually using a range variable so that multiple values can be read or written. These functions operate on files that generally consist of one number per line, that is, numbers separated by carriage returns. *READPRN*, *WRITEPRN*, and *APPENDPRN* are used to read and write whole matrices of data, without using a range variable. We will discuss these functions in the next section. You may find it convenient to use the Windows Notepad or the DOS editor to create data files to be used in your Mathcad documents since both save files as ASCII text without formatting.

The argument of the file-handling functions must be a single filename with no extension or drive or directory name. Mathcad adds the extension .dat to the filename when the function is *READ*, *WRITE*, or *APPEND* and the extension .prn when it is *READPRN*, *WRITEPRN*, or *APPENDPRN*. Mathcad then searches for the file in the current working directory. If the file is not found, the error `file not found` is reported. If the file format is incorrect, the error `file error` is reported.

For example, to read a vector of 100 elements from a file called data1.dat:

Type: `i:0;99`(ENTER)

Type: `X[i:READ(data1)`(ENTER)

To multiply the 100 elements by 2 and then write them back into data1.dat,

type: `WRITE(data1):X*2`(ENTER)

To read or write files that are not in the current directory, you must associate the filename with its directory using the **Associate Filename** command under the **File** menu. When **Associate Filename** is selected, an Associate Filename dialog box appears. In the dialog box, you specify a Mathcad variable, that is, the desired filename without the implied extension, and then tell Mathcad the directory in which the file resides. Recall that Mathcad variables are case sensitive; for example, the file mydata.dat is not the same file as Mydata.dat.

Each Mathcad equation that accesses a data file reopens that file. Therefore, *READ* and *READPRN* both read from the start of a file each time they are used; you cannot use one *READ* command to access the first part of a file and another *READ* command to access the remainder of the file.

WRITE and *WRITEPRN* store data starting at the beginning of the file they name. If *WRITE* or *WRITEPRN* is used two or more times naming the same file as the argument, each execution of the function will overwrite the data written by the previous execution. If you want to add data to an existing file, you must use either *APPEND* or *APPENDPRN*. For example, to append the scalar 5 to the end of the file mydata.dat, you would type

`APPEND(mydata):5`(ENTER)

Note that the *WRITE*, *APPEND*, *WRITEPRN*, and *APPENDPRN* functions must appear alone to the left of an equation. When the length of a data file is not known, you should create a separate file that contains a single value, the length of the data file. In your Mathcad document, you would read the length (an integer) from the file into a variable such as *N*, and then use *N* as a range variable to read the values in the actual data file. Example 3-10 demonstrates this process and the general use of *READ*, *READPRN*, and *WRITE*.

EXAMPLE 3-10

Reading and Writing Vector Data in Data Files

Create a 4-element vector that contains the numbers 1.1, 2.2, 3.3, 4.4. Store the vector in a file called avector.dat. Store the length of the vector, 4, in a separate file called length.dat. Read length.dat and use its contents to correctly read avector.dat. List the contents of avector.dat.

SOLUTION

To create the vector containing the numbers 1.1, 2.2, 3.3, and 4.4, start by defining a range variable that varies from 0 to 3, the vector indices.

Type: `i:0;3`(ENTER)

To create the vector,

type: `V1[i:1.1+1.1*i`(ENTER)

To save the vector length,

type: `WRITE(length):4`(ENTER)

The file length.dat was created when you pressed (ENTER). Alternatively, you could have set a variable, say *N*, equal to 4 and then written the value of the variable.

Type: `WRITE(length):N`(ENTER)

Using a variable is necessary when the actual length of the file value is unknown.

To save the vector in avector.dat,

type: `WRITE(avector):V1`(ENTER)

You could have saved a step by typing:

`WRITE(avector):1.1+1.1*i`(ENTER).

To read and display the contents of length.dat:

Type: `N:READ(length)`(ENTER)

Type: `j:0;N-1`(ENTER)

This range variable, with values from 0 to 3 because $N = 4$, will be used as the index for the following vector READ.

Type: `V2[j:READ(avector)`(ENTER)

You could have given the vector **V2** units by multiplying the READ by the units you desire. For example, you could have typed: `V2[j:READ(avector)*sec`(ENTER).

Type: V2=(ENTER)

Your screen should now show

$i := 0..3$

$V1_i := 1.1 + 1.1 \cdot i$

$\mathrm{WRITE}(\mathrm{length}) := 4$

$\mathrm{WRITE}(\mathrm{avector}) := V1$

$N := \mathrm{READ}(\mathrm{length})$

$j := 0..N - 1$

$V2_j := \mathrm{READ}(\mathrm{avector})$

$$V2 = \begin{bmatrix} 1.1 \\ 2.2 \\ 3.3 \\ 4.4 \end{bmatrix}$$

Reading and Writing Matrices in Data Files

READPRN, *WRITEPRN*, and *APPENDPRN* are used to read and write whole matrices of data. Unlike the case when reading and writing vectors, a range variable is not used. These functions operate on files that generally consist of rows of data, corresponding to the rows of the matrix stored. Each row consists of a matrix row of numbers that are separated by spaces. Each row is terminated by a carriage return.

The argument of the *READPRN*, *WRITEPRN*, and *APPENDPRN* functions must be a single filename with no extension, drive, or directory name. Mathcad adds the extension .prn to the argument. Data files created outside Mathcad must be given the .prn extension.

EXAMPLE 3-11

Reading and Writing Matrix Data in Data Files

Create the following 4 × 3 matrix:

0.0	0.1	0.2
1.0	1.1	1.2
2.0	2.1	2.2
3.0	3.1	3.2

Store the matrix in a file called matrix.prn. Read the file contents and list them.

SOLUTION

To create the matrix:

Without pressing (ENTER), type: `M1:`

Select Matrices under the Math menu.

In the Matrices dialog box, set Rows to 4 and Columns to 3

Click on the Create button in the Matrices dialog box.

Click on the upper left-hand placeholder and, using (TAB), enter the specified element values.

To store the matrix **M1**, read it back into matrix **M2**; then display **M2**.

Type: `WRITEPRN(matrix):M1`(ENTER)

Type: `M2:READPRN(matrix)`(ENTER)

Type: `M2=`(ENTER)

You may want to press (CTRL)+s to separate the regions. Your screen should now show

$$M1 := \begin{bmatrix} 0.0 & 0.1 & 0.2 \\ 1.0 & 1.1 & 1.2 \\ 2.0 & 2.1 & 2.2 \\ 3.0 & 3.1 & 3.2 \end{bmatrix}$$

WRITEPRN(matrix) := M1

M2 := READPRN (matrix)

$$M2 = \begin{vmatrix} 0 & 0.1 & 0.2 \\ 1 & 1.1 & 1.2 \\ 2 & 2.1 & 2.2 \\ 3 & 3.1 & 3.2 \end{vmatrix}$$

Try It Perform the following Mathcad operations:

$N := 10$

WRITE(filesize) :=

$i := 0..N - 1$

$Y_i := 2 + i \cdot 2$

WRITE(list) := Y_i

- What do these operations do?
- Display the contents of **Y**. Of what do they consist?
- What happens if you change the value of *N*?

3-3 SOLVING MATHEMATICAL PROBLEMS

Mathcad can determine the *root* of any mathematical expression; that is, it can determine the value of a variable in an expression for which the value of the expression is zero. Mathcad also can solve sets of equations, determine the values of summation and product expressions, generate random numbers, and sort arrays. It has built-in statistical functions, can plot histograms of collections of data, and can do a least squares fit on a collection of data. In this section we will explain how to use each of these Mathcad features.

Solving Single Equations

Engineers typically develop functions that describe the systems they design and analyze. Such functions may describe, for example, the nature of a robotic control system, an electronic amplifier, or a mechanical system. The roots of these functions often give important information about the systems they describe, such as their stability.

To determine the roots of a function, use the *root* function. This function takes as its arguments two scalars. The first argument can be a function defined elsewhere in the document; the second is a variable name that appears in the function or expression. The *root* function varies the value of the variable until the function or expression has a value of zero and returns the resulting value of the variable.

You must supply Mathcad an initial "guess" for the value of the root. Mathcad uses this guess as a starting point in its search for a root. In most cases, your choice of a guess value does not make much difference. However, for functions or expressions with multiple roots, your guess will determine which root is found.

The *root* function ceases its search on finding a root. If there are other roots, they will not be found unless you provide guesses that are more appropriate for finding those roots. When in doubt about how many roots there are and where they are, it is best to plot the function in question. (Plotting is the focus of Chapter 4.)

EXAMPLE 3-12

Determining the Root of a Function

Determine the values of x for which the following functions have values of zero:

- $f(x) = 5x + 3$
- $g(x) = 3x^2+x-8$
- $h(x) = 4x^3-e^x$

SOLUTION

Clear the document window.

Type: `f(x):5*x+3`(ENTER)

Type: `g(x):3*x^2`(↑)(↑)`+x-8`(ENTER)

Type: `h(x):4*x^3`(↑)(↑)`-e^x`(ENTER)

Now give Mathcad an initial guess for *x*. The guess normally does not have to be that accurate; let's use zero for all three functions.

Type: `x:0`(ENTER)

To determine the roots of the three functions:

Type: `root(f(x),x)=`(ENTER)
Type: `root(g(x),x)=`(ENTER)
Type: `root(h(x),x)=`(ENTER)

Your screen should now show

$$f(x) := 5 \cdot x + 3$$

$$g(x) := 3 \cdot x^2 + x - 8$$

$$h(x) := 4 \cdot x^3 - e^x$$

$$x := 0$$

$$\text{root}(f(x), x) = -0.6$$

$$\text{root}(g(x), x) = 1.475$$

$$\text{root}(h(x), x) = 0.831$$

You can check to make sure these roots truly result in the value of the function being zero. For example, typing `f(-0.6)=`(ENTER) should return a result of 0.

Some functions create problems for Mathcad. You should attempt to find the roots of the functions $f(x)$ = tan(x)-x and $f(x)$ = sqrt(1-x^2). Plotting the functions should help. See the QuickSheets for more information on solving many types of algebraic problems.

Try It Determine the values of *x* for which the following functions have values of zero. For *g*(*x*), find the two roots nearest *x* = 0.

- $f(x) = -3x + 5$
- $g(x) = cos(377x)$

Complex Numbers

As stated earlier, the roots of functions that describe engineering systems, for example, electrical and mechanical systems, yield information about, among other things, the stability of the systems. The roots of many of these functions are complex; therefore, engineers and their software tools must be able to deal with complex numbers. In Example 3-13, the roots of a function are complex.

EXAMPLE 3-13

Finding Complex Roots

Determine the square root of -4 and all the roots for the function $f(y) = y^2 - y + 5$

SOLUTION

Use a backslash (\) to create a radical and determine the square root of –4.

Type: `\-4=`(ENTER)

Mathcad should have given the answer 2*i*, using *i* to indicate an imaginary value.

Give Mathcad an initial guess of 0 for the root of $f(y) = y^2 - y + 5$.

Type: `y:0`(ENTER)

Now have Mathcad compute the root of $f(y) = y^2 - y + 5$.

Type: `root(y^2`(↑)(↑)`-y+5,y)=`(ENTER)

Mathcad should have given you a `not converging` error message. It is looking for a real root because you gave it a real initial guess. However, the function has no real roots, so you must instruct Mathcad to search for an imaginary root, as follows:

Click on the 0 in your initial guess for y.

Delete the 0 and type: `1i`(ENTER)

Mathcad does not show the 1. Note that it is `1i` and not `1*I` that you entered. Had you entered 1**i*, Mathcad would have interpreted the *i* as a variable name. This would have also been so if you had simply entered *i* by itself.

With some appropriate text added, you should now see

$\sqrt{-4} = 2i$

$y := i$ This is an initial guess

$\mathrm{root}(y^2 - y + 5, y) = 0.5 + 2.179i$

This function has another root. Find it by again editing the initial guess for y to be -1*i*. Your screen should now show the other root, the complex conjugate of the first root.

$\sqrt{-4} = 2i$

$y := -i$ This is an initial guess

$\mathrm{root}(y^2 - y + 5, y) = 0.5 - 2.179i$

Try It Determine all of the roots of the following functions.

- $g(w) = 3w^2 - w + 18$
- $h(w) = w^4 - 16$

Solving Sets of Equations

With Mathcad, you can solve a set of up to 50 simultaneous equations. The *find(x,y)* function is used to find the points of intersection of the two functions, *x* and *y*. Four steps are involved in solving a set of equations.

1. Provide an initial guess for each of the variables for which you intend to find values. As when finding roots, Mathcad solves a set of equations by making a series of guesses, starting with the guess you provide. For example, an initial guess of 0,0 as the intersection of two functions might be entered by typing

   `x:0(ENTER)y:0(ENTER)`

2. Type `given(ENTER)` to tell Mathcad that what follows is a set of equations. The word *given* is case insensitive.
3. Type the equations and inequalities in any order below the word *given*. When typing an equation, you must use (CTRL)= rather than just = or : as you have in other situations. When typing an inequality, use < for less than, > for greater than, (CTRL)+`0` for greater than or equal, and (CTRL)+9 for less than or equal. Mathcad does not allow the use of the not equal constraint. For example, the following would be appropriate when determining the point of intersection of two lines:

   ```
   given(ENTER)
   y(CTRL)=-6*x+3(ENTER)
   y(CTRL)=5*x-2(ENTER)
   ```

4. Type any expression that involves the *find* function. For example, you could type `find(x,y)=` to determine the values of *x* and *y* that solve two simultaneous equations involving *x* and *y*.

   `find(x,y)=(ENTER)`

The entries made in steps 2, 3, and 4 form a *solve block.* When you use this four-step process, you are using a solve block to solve simultaneous equations.

EXAMPLE 3-14

Solving Sets of Equations

Use a solve block to determine where the parabola y = $1/2x^2-8$ intersects the circle $x^2+y^2=25$.

SOLUTION

If you plot these two equations, you will notice that the parabola intersects the circle in four places, once in each of the four quadrants of the X-Y plane. Which intersection you find will be determined by the initial guess you provide for *x* and *y*.

To find the intersection in the first quadrant, provide the guess:

Type: `x:1`[ENTER]`y:1`[ENTER]

To start the solve block:

Type: `given`[ENTER]

Now type the two equations, remembering to hold the [CTRL] key down while pressing the equals sign (=):

Type: `y`[CTRL]`=1/2`↑↑`*x^2`↑↑`-8`[ENTER]

Type: `x^2`↑↑`+y^2`↑↑[CTRL]`=25`

Completing the solve block, type: `find(x,y)=`[ENTER]

The screen shows

$$x := 1$$

$$y := 1$$

given

$$y = \frac{1}{2}\cdot x^2 - 8$$

$$x^2 + y^2 = 25$$

$$\text{find}(x, y) = \begin{pmatrix} 4.508 \\ 2.162 \end{pmatrix}$$

To find the remaining intersections, you must go back and edit your initial guess. In this example, you must merely provide a guess in each of the four quadrants to find the intersection in the corresponding quadrant.

Note that Mathcad provided the result in the preceding example in the form of a vector in which the answers are listed top to bottom, corresponding to the arguments of the *find* function listed left to right. In this case, the value of *x* was listed first in the vector, followed by the value of *y*.

You do not have to use the *find* function in the way it was used in the example. You could have set the function equal to another variable, as in

$$z := \text{find}(x, y)$$

$$z = \begin{pmatrix} 4.508 \\ 2.162 \end{pmatrix}$$

Try It

Use a solve block to solve the following set of equations:

- $3x+4y-2z=1$
- $-2x-z=0$
- $x-6y-2z=-2$

Use a solve block, rather than the *root* function, to determine where the parabola $y = x^2-16$ intersects the positive X-axis. Your solve block will have three equations: $y = x^2-16$, $y = 0$, and $x > 0$.

Summations and Products

The summation operator sums an expression over all values of an index, whereas the product operator multiplies an expression over all values of an index. These operators often are used in computing the values of infinite series of sums or products that appear in many engineering disciplines.

To perform a summation operation, type CTRL+SHFT+4. To perform a product operation, type CTRL+SHFT+3. In either case, the appropriate operator symbol will appear at the red crosshair. You then enter the values of your choice in the placeholders that appear with the symbol.

EXAMPLE 3-15

Determining Summations and Products

Determine the value of each of the following series:

- $1 + 2 + 3 + \ldots + 20$
- $10 * 9 * 8 * \ldots * 1$
- $(1^2 + 2^2 + 3^2 + \ldots N^2) / N$ for $N = 20$

SOLUTION

To compute the sum $1 + 2 + 3 + \ldots +20$,

type (or click on the Calculus Palette): CTRL+SHFT+4

You should now see:

$$\sum_{\blacksquare = \blacksquare}^{\blacksquare} \blacksquare$$

Click on the placeholder to the left of the equals sign (=) and type: `i`

Click on the placeholder to the right of the equals sign (=) and type: `1`

Click on the placeholder above the summation symbol and type: `20`

Click on the placeholder to the right of the summation symbol.

Type: `i=`ENTER

You should now see

$$\sum_{i=1}^{20} i = 210$$

To compute the product $10 * 9 * 8 * \ldots * 1$:

Type: CTRL+SHFT+3

Click on the placeholder to the left of the equals sign (=) and type: `p`

Click on the placeholder to the right of the equals sign (=) and type: `10`

Click on the placeholder above the summation symbol and type: `1`

Click on the placeholder to the right of the summation symbol.

Type: `p=`(ENTER)

You should now see

$$\prod_{p=10}^{1} p = 3.629 \cdot 10^{6}$$

To compute the sum ($1^2 + 2^2 + 3^3 + \ldots N^2$) / N for $N = 20$:

Type: `N:20`(ENTER)

Type: `1/N`(↑)(↑)`*`(CTRL)+(SHFT)+4

Click on the placeholder to the left of the equals sign (=) and type: `n`

Click on the placeholder to the right of the equals sign (=) and type: `1`

Click on the placeholder above the summation symbol and type: `N`

Click on the placeholder to the right of the summation symbol.

Type: `n^2=`(ENTER)

You should see

$$N := 20$$

$$\frac{1}{N} \cdot \sum_{n=1}^{N} n^2 = 143.5$$

Try It Determine the values of the following expressions:

$$\frac{1}{2 \cdot \pi} \cdot \left[\sum_{j=1}^{15} \left(\cos(j)^2\right) \right]$$

$$\prod_{j=1}^{12} \sqrt{j}$$

Random Numbers

Mathcad includes a function *rnd*(*x*) that returns a random real number between 0 and *x*, but not including 0 or *x*. To generate a random integer, you must use Mathcad's *floor* function to convert the random real number into an integer. The function *floor*(*x*) returns the greatest integer that is less than or equal to *x*. For example, if *rnd*(*10*) returns the real number 0.571, *floor*(*rnd*(*10*)) will return the integer 0. Since *rnd*(*10*) will never return 10, the largest integer *floor*(*rnd*(*10*)) can return is 9.

We will use the *rnd* function throughout the remainder of this section to generate data sets of random numbers. Just as you might do in your engineering work, vectors and matrices of random numbers are used in this book to test mathematical algorithms. Generating a vector of random integers is a three-step process:

1. Define a range variable.
2. Define the vector.
3. Test the vector contents to make sure they are what you wanted.

Example 3-16 demonstrates how to generate a vector of random numbers.

EXAMPLE 3-16

Generating a Vector of Random Numbers

Create a vector that contains 10 random integers between 100 and 199.

SOLUTION

First, create the range variable.

Type: `i:0;9`(ENTER)

Using [to indicate that *i* is a subscript, define the vector.

Type: `X[i:100+floor(rnd(100))`(ENTER)

The algorithm for obtaining the sets of random numbers you want is rarely obvious. You must determine these algorithms on your own. In this case, the expression *floor*(*rnd*(*100*)) generates integers from 0 to 99, which are then added to the integer 100 in order to generate what we want—integers from 100 to 199.

To test the contents of the vector **X**:

Type: `X[i=`(ENTER)

Click to the right of the resulting table.

Type: `min(X)=`(ENTER)`max(X)=`(ENTER)

Since you are dealing with random numbers here, the values that appear on your screen may differ; however, the result of the work is

$i := 0..9$

$X_i := 100 + \text{floor}(\text{rnd}(100))$

X_i

103
164
108
194
127
198
129
136
121
124

$\min(X) = 103$

$\max(X) = 198$

Note that Mathcad will always generate the same sequence of random numbers each time Mathcad is used. To obtain a different sequence, you must change the seed number used by the random number generator. To provide Mathcad with a new seed, select **Randomize...** under the **Math** menu.

Try It Create a vector that contains 20 random real numbers from -100 to +100.

Sorting

Mathcad provides a function for sorting vectors and matrices. The function *sort*(**v**) returns vector **v** sorted in ascending order. The function *csort*(***M***,*n*) returns the matrix **M** with its rows sorted in ascending order according to the values of the elements in column n. The function *rsort*(***M***,*n*) returns the matrix **M** with its columns sorted in ascending order according to the values of the elements in row *n*.

Mathcad also provides a function *reverse* that accepts either a vector or a matrix. This function reverses the order of the elements in the argument passed to it. Therefore, if you want to sort a vector or matrix in *descending* order, first sort it in ascending order and use reverse to *reverse* the order.

Each of these array functions accepts vectors and matrices that include complex numbers; however, Mathcad ignores the imaginary parts while sorting.

EXAMPLE 3-17

Sorting a Matrix

Create a 4×5 matrix of random integers from 0 to 499. Sort the matrix into descending order according to the elements in the leftmost column.

SOLUTION

To create a matrix of random numbers requires that we use range variables. Using the Matrices dialog box is not an option since it requires the user to enter the element values manually, values that must be random. Defining each element individually would take too long, especially when the matrix is particularly large. For example, you could start with the upper left element in the matrix.

Type: `X[0,0:floor(rnd(500))`(ENTER)

And then type: `X[0,0=`(ENTER)

You might see something like

$X_{0,0} := \text{floor}(\text{rnd}(500))$

$X_{0,0} = 96$

Using range variables to define a matrix requires two indices.

Type: `i:0;3Wj:0;4`(ENTER)

Now create the matrix.

Type: `X[i,j:floor(rnd(500))`(ENTER)

To view the original matrix,

type: X=(ENTER)

To sort the rows of the matrix according to column 0, reverse the row order, and display the result:

type: `reverse(csort(X,0))=`(ENTER)

You may need to press (CTRL)+s to separate the regions.

Although your random numbers will be different, the screen should show something like

$i := 0..3$

$j := 0..4$

$X_{i,j} := \text{floor}(\text{rnd}(500))$

$$X = \begin{bmatrix} 0 & 96 & 292 & 175 & 411 \\ 87 & 355 & 151 & 45 & 73 \\ 494 & 59 & 4 & 265 & 300 \\ 83 & 225 & 28 & 391 & 259 \end{bmatrix}$$

+

$$\text{reverse}(\text{csort}(X,0)) = \begin{bmatrix} 494 & 59 & 4 & 265 & 300 \\ 87 & 355 & 151 & 45 & 73 \\ 83 & 225 & 28 & 391 & 259 \\ 0 & 96 & 292 & 175 & 411 \end{bmatrix}$$

Note that column 0 is in reverse order. During the sorting process, the remainder of each row was moved with that row's element in the leftmost column.

Try It

Create a 6 × 3 matrix of random real numbers from 100 to 200. Sort the matrix into ascending order according to the elements in the rightmost column.

Histograms

A *histogram* displays the frequency with which a number or set of numbers appears within a data set. For example, if a petroleum engineer had a set of data on the production of a large number of oil wells, he or she might want to know how many wells were producing between 100 and 200 barrels of oil per day.

Mathcad provides the *hist*(**int**,**v**) function for creating histograms. This function returns a vector that represents the frequencies with which values in the vector **v** fall in the intervals defined by the vector **int**. Both function arguments contain real numbers. The returned vector contains integers and one less element than does **v**. Example 3-18 demonstrates the process of creating a histogram.

EXAMPLE 3-18

Generating a Histogram

Create a 100-element vector of random real numbers that range from -50 to +50 kilograms. Then determine how many of these numbers lie between -50 and -25, -25 and 0, 0 and +25, and +25 and +50.

SOLUTION

You may want to clear the screen before proceeding. Using a range variable to create the 100-element vector, **V**,

type: `i:0;99` (ENTER)

To create 100 random numbers between, but not including, -50 and +50,

type: `V[i:(-50+rnd(100))*kg` (ENTER)

Now use a range variable to define the interval vector, **INT**.

Type: `j:0;4` (ENTER)

Type: `INT[j:(-50+25*j)*kg` (ENTER)

Note that, since *j* varied from 0 to 4, the interval vector has 5 elements; therefore, the result returned by *hist*(INT,V) will be a 4-element vector. There are four groupings of numbers in which we are interested. Also note that selecting **Matrices** under the **Math** menu and using the Matrices dialog box to define this vector would have been an equally valid approach.

To create and display the histogram,

type: `hist(INT,V)=` (ENTER)

Although the solutions to this problem will vary because you are using random numbers, you should see something like

```
i := 0..99

V_i := (-50 + rnd(100))·kg

j := 0..4

INT_j := (-50 + 25·j)·kg

                 | 25 |   <== no. of no.s >= -50 and < -25
hist(INT, V) =   | 20 |   <== no. of no.s >= -25 and < 0
                 | 31 |   <== no. of no.s >=0 and < +25
                 | 24 |   <== no. of no.s >= +25 and < +50
```

The histogram has been annotated appropriately. When you wish to include text, remember that you just click the mouse at the location where you want the text, type a double quote ("), and then enter the text. To quit entering text, click the mouse outside the text region.

Note that, if the numbers within the vector are truly random, probability theory dictates that the four groups in the preceding histogram should contain roughly the same numbers of numbers. However, there is no guarantee that this will be so when dealing with a finite set of random numbers. One hundred numbers is not a large set; had 1,000 numbers been generated, the four groups likely would have been more similar in size.

Most engineers prefer to see histograms in the form of a bar chart. In Chapter 4, we will describe how to plot bar chart histograms.

Try It Create a 200-element vector of random real numbers that range from -100 to 0. Then determine how many of these numbers lie between -100 and -50, -50 and -25, -25 and -10, and -10 and 0. You will have to use the Matrices dialog box to create the interval vector.

Statistics

Engineers often deal with large amounts of data and want to know the average, standard deviation, and variance of that data. Mathcad offers all three as the built-in functions *mean*(**v**), *stdev*(**v**), and *var*(**v**), where **v** is a vector. Example 3-19 demonstrates how to use the three functions.

EXAMPLE 3-19

Using Statistical Functions

Determine the mean, standard deviation, and variance of the following data: 123, 83, 94, 87, 122, 133, 100.

SOLUTION

Large quantities of data are best handled using data files, a subject covered in Section 3-2. Here, however, you have only seven data items, so you can easily create a vector and then fill it manually.

Clear the screen and, without typing (ENTER), type: `D:`

Select Matrices under the Math menu.

The **Matrices** dialog box should appear.

Specify 7 rows and 1 column in the **Matrices** dialog box and click on the Create button.

Click on the placeholders in the vector **D**, entering the values 123, 83, 94, 87, 122, 133, 100.

Press: (ENTER)

Type: `mean(D)=`(ENTER)`stdev(D)=`(ENTER)`var(D)=`(ENTER)

You should now see

$$D := \begin{bmatrix} 123 \\ 83 \\ 94 \\ 87 \\ 122 \\ 133 \\ 100 \end{bmatrix}$$

$$\mathrm{mean}(D) = 106$$

$$\mathrm{stdev}(D) = 18.299$$

$$\mathrm{var}(D) = 334.857$$

The QuickSheets offer much more information on performing statistical analysis.

Try It Determine the mean, standard deviation, and variance of the integers 1, 2, 3, ..., 12.

Application 2

COMPUTER NETWORKS

The design and maintenance of local area networks (LANs) are common tasks for computer engineers. One aspect of the performance of a network is the utility of the cable links that interconnect the computers in the network. Utility is a measure of the percentage of time that a cable carries binary bits of data. The greater the utility, the better the performance of the network. All data in most networks are broken up into groups of bits called packets, often of a fixed size. The utility of a cable link is defined by the equation

$$U = 1 \,/\, (1+a)$$

where $a = tt \,/\, pt$. The variable tt is the transmission time, that is, the time it takes a source computer to put a packet of data on a cable link. The propagation time pt is the time it takes the packet to reach the destination computer at the other end of the link.

1. Define the Problem

Determine the average of the utilities of the 10 cable links that compose a computer network.

2. Gather Information

Assuming the packet size is 1024 bits and the transmission rate is 1 Mbps (millions of bits per second), where 1 M equals 1024^2, we have a packet transmission time *tt* of 1024 / 1024^2, or 0.98 millisecond. The propagation time *pt* depends on the length of the cable, and the propagation speed of bits (voltage or light pulses) on the cable. Readily available test equipment measures the propagation time of cables. Let's assume we have measured these times for the 10 links in our network and have stored the data in a data file called trantime.dat. The following, in seconds, are the data stored in trantime.dat:

9.8E–4,15E–4,7E–4,10.2E–4,20E–4,10E–4,9E–4,10E–4,10E–4,9.7E–4

Note that Mathcad accepts data written in exponential form, as shown here.

3. Generate and Evaluate Potential Solutions

We need to read the transmission times stored in trantime.dat and from that data create a vector containing the utility values for the 10 links. Lastly, we must compute the average of those utilities.

4. Refine and Implement a Solution

The following Mathcad work implements the solution:

$i := 0..9$	The vectors' index
$pt_i := \mathrm{READ}(\mathrm{trantime}) \cdot \mathrm{sec}$	Read the propagation times
$tt := 0.98 \cdot 10^{-3} \cdot \mathrm{sec}$	The transmission time is 0.98 ms
$a_i := \dfrac{tt}{pt_i}$	Create a vector of values of a=tt/pt
$U_i := \dfrac{1}{1 + a_i}$	Create a vector of values of utility
$Uave := \dfrac{1}{10} \cdot \displaystyle\sum_{i=0}^{9} U_i$	Compute the average utility
$Uave = 0.519$	The links are utilized, on the average, 51.9 percent of the time

Therefore, we have an average utility of 0.519; that is, the links are being utilized, on average, about 52 percent of the time. This is a typical value for networks that have relatively long cable links.

5. Verify and Test the Solution

There are a number of ways of measuring computer network performance. Many hardware network analyzers are available from various manufacturers. Software tools also are available. It is the corroboration of a collection of data that lets an engineer know that the overall picture is correct.

What If

What if the engineer wants to determine which link has the highest utility and which link has the lowest utility?

SUMMARY

This chapter began by showing you how to use Mathcad's built-in functions and how to create and use your own mathematical functions. A description of how to use units, such as meters and seconds, in your equations followed. We then presented an explanation of how to perform iterative tasks using range variables and how to use range variables to create tables of data. In the second section, we described how to create, edit, and compute vectors and matrices. We listed some of the available array operators and functions, and explained how to read and write data files. In the final section of this chapter, we provided many examples of how to use Mathcad's more popular features. These examples included the solution of a single equation and a set of equations, the computation of summations, the generation of random numbers, data sorting, histogram creation, and the use of statistical functions. Throughout the chapter, you were reminded about Mathcad's extensive examples and explanations to be found in its QuickSheets and Electronic Books. The Electronic Books also offer detailed help in such specialized applications as mechanics, electromagnetics, and the properties of liquids, solids, and metals.

Key Words

data file
histogram
matrix
range variable
root
scalar
solve block
units
vector

Problems

1. Determine the value of the expression $5e^{-t}(3\cos(\pi/8) + 4\sin(\pi/8))$ at $t = 1$.
2. Graph the function $5e^{-t}(3\cos(\pi/8) + 4\sin(\pi/8))$ over the range $t = 0$ to 5.
3. Using appropriate units, determine the necessary acceleration of a rocket car that must complete a quarter-mile drag race in less than 5 seconds from a standing start. Assume the acceleration is constant throughout the entire race. Assume all other parameters, such as wind resistance, are negligible. Give your answer in feet/second^2 and in g's.
4. Create two tables that display, respectively, the time in seconds and

position in feet of the rocket car in problem 2 for $t = 0$ to $t = 5$ seconds in 0.1 second increments. Format the position table so that the displayed precision is 0, that is, so that the values in the table are rounded to whole numbers.

5. Create a 10-element vector in which each element is the square of its index. That is, the elements are 0, 1, 4, ..., 81. Graph the vector.
6. Create a 4×3 matrix **X**, in which the values of the elements are 1 more than the row number in which they reside. Therefore, the first row will contain all 1s, and the last row will contain all 4s.
7. Determine the square of the following matrix and use built-in functions to determine the minimum and maximum values in the resulting matrix:

X =	7	0	1	-3
	0	1	8	5
	-1	2	7	4
	-2	9	8	3

8. Create the following 4×3 matrix, store it in a data file, read the file back into a matrix **M2**, and create a third matrix **M3** that is the sum of **M1** and **M2**. Display **M3**.

M1 =	0.0	0.3	0.6
	3.0	3.3	3.6
	6.0	6.3	6.6
	9.0	9.3	9.6

9. Save the vector **V1** in a data file, append **V2** to that data file, read the file into a vector **V3**, and display **V3**.

V1 =	1.1	**V2** =	4.4
	2.2		3.3
	3.3		2.2
	4.4		1.1

10. Find the roots of the following functions. For $g(x)$, find the two roots nearest $x = 0$. For $h(x)$, find both roots.
    a. $f(x) = -5x + 3$
    b. $g(t) = 1.707\cos(2\pi 377t)$
    c. $h(x) = 2x^2 + x - 12$
11. Graph each of the functions listed in Problem 10 and verify the location of the roots of each function.
12. Find all (complex) roots of each of the following functions:
    a. $f(x) = 2x^2 + 12$
    b. $g(x) = 4x^2 - 4x + 24$
    c. $h(x) = 3x^2 - x + 18$

13. Use a solve block to determine the following:
    a. where the parabola $y = x^2 + x - 4$ intersects the line $y = 5$.
    b. where the parabola $y = (3/4)x^2 - 16$ intersects the line $y = 5x + 5$.
14. Use a solve block to solve the following set of equations:
    a. $4x + 4y - 4z = 0$
    b. $-4x - 2z = 0$
    c. $x - y - z = -2$
15. Determine the value of each of the following series.
    a. 1 + 2 + 3 + ... + 200
    b. 30 * 9 * 8 * ... * 1
    c. $(1^3 + 2^3 + 3^3 + \ldots N^3) / N$ for $N = 10$
16. Determine the value of the following expressions:
    a.
    $$\frac{16}{\pi}\cdot\sum_{n=1}^{1000}\frac{\sin\left[(2\cdot n - 1)\cdot\frac{\pi}{2}\right]}{2\cdot n - 1}$$
    b.
    $$\frac{1}{100}\cdot\prod_{N=1}^{10}\prod_{M=10}^{1}\sqrt{N - M}$$

Use the five-step problem-solving process while working the remaining problems.

17. Assume a group of 10 manufactured parts must be measured for length, width, and height. The parts are supposed to be cubic, 10 cm on each side. Create a 10 × 3 matrix of random integers from –9.8 to +10.2. The first column of the matrix holds length values, the second width values, and the third height values. Sort the matrix into descending order according to the width values and display the result. Determine the volumes of the smallest and the largest cube.
18. Assume you are an automotive engineer who is collecting horsepower figures on 20 automobile engines. Create a 20 × 2 matrix. Use a range variable to store the numbers 1,2,3, ... , 20, representing the engine numbers, in the first column. Use a range variable to store random integers between 140 and 150, representing the horsepower of the 20 engines, in the second column. Sort the matrix in descending order according to horsepower. Display the result. Determine and display the numbers of the engines that have the least and most power, along with the corresponding power for both engines, using the correct units (hp). Determine and display in the correct units the mean power of the 20 engines.
19. Assume you are a petroleum engineer who must determine how many of 50 oil wells are producing fewer than 10 barrels of oil per day, how many are producing between 10 and 20 barrels, 20 to 30, ... , 90 to 100. Create a 50-element vector of random integers that range from 0 to 100, representing the number of barrels of oil produced per day for each of the 50 wells. Then determine how many of the random values lie between 0 and 10, 10 and 20, 20 and 30, ... , 90 and 100.

20. Assume you are an electrical engineer who must determine the total kilowatt-hours consumed by five homes during one week when the five homes' average power consumption during that week is 1.1KW, 1.3KW, 0.9KW, 1.4KW, and 1.3KW. Use all correct units.
21. Assume a group of 50 resistors have been manufactured to have a resistance of 2.5 Kohm with a 10 percent tolerance. The resistances are measured automatically and stored in a data file ohms.prn that consists of 5 lines of data with 10 values stored in each line. Create the data file, putting random values from 2,200 to 2,800 in it. Determine the average of the resistances stored in ohms.prn. Use a histogram to show how many of the resistances are out of tolerance.
22. Assume a class of 10 students who have taken 5 tests during a semester. Create a 10×5 matrix and fill it with random numbers from 70 to 100. Create a vector that contains the test average for each of the 10 students. Assuming that students with a test average from 70 to 80 get a C, from 80 to 90 get a B, and from 90 to 100 get an A, create a histogram that shows the number of students who received Cs, Bs, and As.
23. Assume a class of 10 students that has taken 5 tests during a semester. Create a 10×5 matrix and fill it with random numbers from 50 to 100. Create a vector that contains the average of each of the 5 tests. Create another vector that contains the test average for each of the 10 students. Determine the average of all the tests. Determine the lowest and highest scores made by the students. Determine the mean, standard deviation, and variance of the students' scores on each test.

4 Engineering Charts and Plots

Surveying Methods

The practice of surveying has existed ever since land was allocated to certain owners and earth had to be moved to construct structures. The Babylonians, about 3500 B.C., produced maps indicating that they had used surveying to obtain field information. The Egyptians also used surveying methods as early as 1400 B.C. to reestablish boundary markers that the flood waters of the Nile had obliterated.

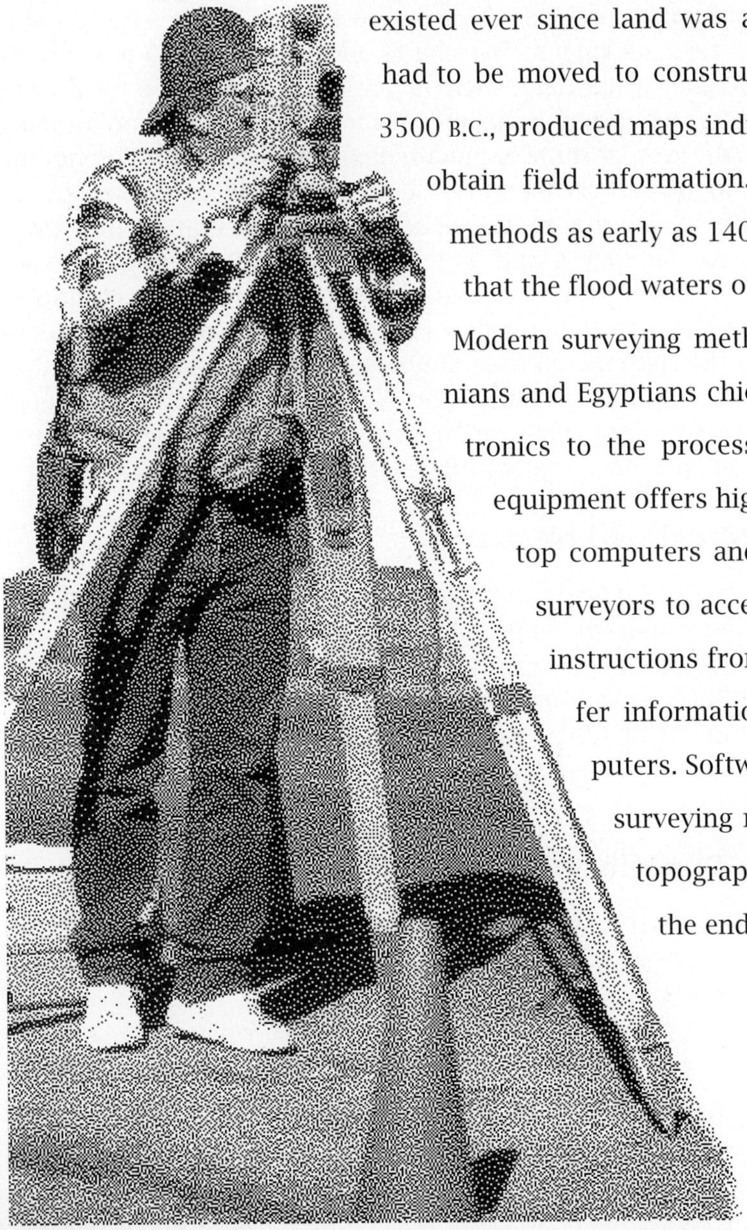

Modern surveying methods differ from those of the Babylonians and Egyptians chiefly because of the application of electronics to the process. For example, laser-based surveying equipment offers high degrees of accuracy. In addition, laptop computers and cellular phone communication allow surveyors to access complete project specifications and instructions from home office computers and to transfer information from the field back to those computers. Software tools have simplified and advanced surveying methods. Mathcad is used to develop a topographical map in the application problem at the end of this chapter.

INTRODUCTION

In this chapter, you will learn to create graphs and plots. The chapter begins with the subject of graphing functions and vectors, or one-dimensional arrays. The creation of histograms is the next topic, followed by the computation and graphing of a least squares fit of data. We then discuss the drawing of polar plots. The remaining two plotting subjects covered in this chapter are surface plots and contour plots. The chapter concludes with an application problem from the field of civil engineering in which we describe the creation of a topographical survey map.

4-1 GRAPHS

As you have seen in previous chapters, graphs are easy to create using Mathcad. Typically, you start by defining a range variable that will act as the independent variable of the graph, that is, the variable plotted along the X-axis. The data to be graphed can be specified by a function that is defined directly in the graph to the left of the Y-axis, or it can be defined above the graph as a user-defined function, or it can be stored in a vector. You can plot multiple functions or vectors on a single graph. You can plot data in various formats such as bar charts and histograms. You can perform automatic analysis of data, such as doing a least squares fit of a set of X,Y data points to determine the linear function that best approximates the data. You can even plot one function against another, producing, for example, a Lissajous pattern.

All the graphs you create in this section use a *Cartesian* set of coordinates, that is, the familiar X and Y axes. In Mathcad, these graphs are called *X-Y plots.* Polar plots are covered in Section 4-2. For many examples of the creation of Mathcad's many types of graphs and plots, select **Quick-Sheets...** under the **Help** menu; then double click on Graphing. Remember that the type of graph or plot that you wish to create can be selected either under the **Graphics** menu or by selecting among the options presented when you click on the Graphing Palette button on the palette.

Graphing Multiple Functions

Although you graphed a few simple functions in earlier chapters, in this chapter the subject of graphing is covered in more detail. In the following examples, you will review several ways of plotting functions and learn how to plot multiple functions. In Example 4-1, you again plot a simple function, this time using units to gain some practice using this important Mathcad feature.

EXAMPLE 4-1

Graphing a Function

Graph the function $y(t) = 5000 - 0.5at^2$, where a equals 9.8 m/sec^2, the acceleration due to gravity; t is in seconds; and $y(t)$ is in meters. Let t vary from 0 to 20 seconds.

SOLUTION

If your document window is not already cleared, clear it now.

Define the independent variable t by typing: `t:0*sec;20*sec` (ENTER)

Define the function by typing: `f(t):5000*m-0.5*9.8*m/sec^2` (↑)(↑)(↑) `*t^2` (ENTER)

Select **Create X-Y Plot** under the **Graphics** menu (or use the Graphics Palette button on palette).

A box should appear with placeholders along the bottom (the X-axis) and left-hand side (the Y-axis).

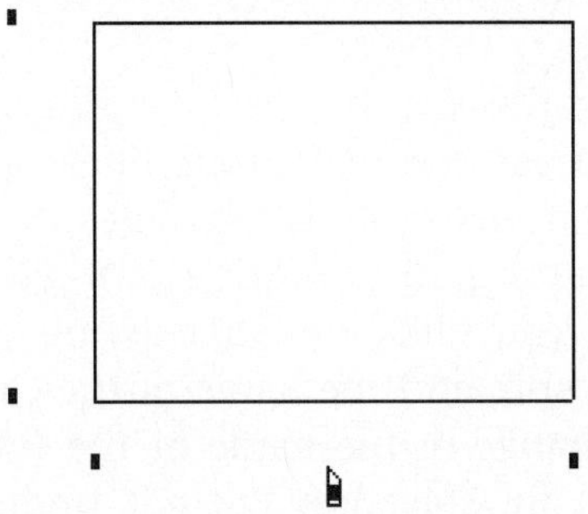

A selection box (discussed in Section 2-4) appears around the placeholder below the center of the X-axis because Mathcad expects you to enter the independent variable first.

Designate the independent variable in the selection box by typing: `t`

Click on the placeholder to the left of the center of the Y-axis.

Type: `f(t)` (ENTER)

You should now see

$t := 0 \cdot sec .. 20 \cdot sec$

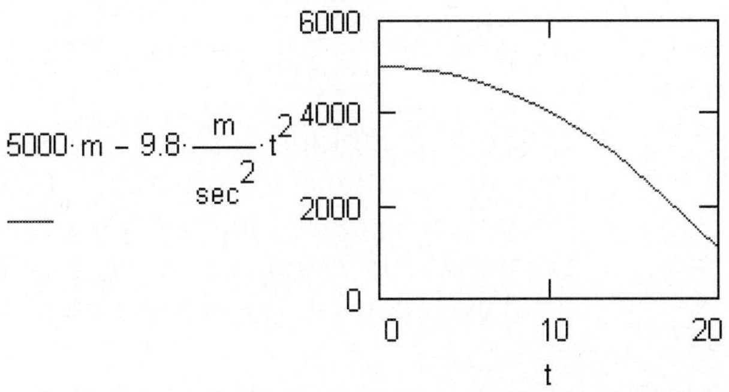

Note that you did not give units to *t* in the function definition; this is because Mathcad already knew its units from the declaration of the range of *t* above the graph. Mathcad automatically set the range on the Y-axis for you, based on the fact that you defined the X-axis range to be from 0 to 20.

The line under f(t) to the left of the graph indicates the line type used to plot the function so that it can be differentiated from other plotted functions. In this case, the line type was a solid line; other types are dashed lines and dotted lines. If more than one function is plotted in a graph, each function definition is listed to the left of the Y-axis, underlined with a small example of the line type used to plot that function in the graph. For example, the first function is always plotted with a solid line, whereas the second is plotted with a dashed line. Example 4-2 demonstrates the process.

Remember, the units you use in Mathcad are not merely labels. Mathcad "understands" the relationships between units of mass, distance, time, and so on. Therefore, had you entered 5000*sec instead of 5000*m in the graph function, Mathcad would have given you an `incompatible units` error message.

Mathcad graphs can be formatted in a number of different ways. The defaults for most graphs include linear axes scales, automatic axes scaling, and no options such as grid lines or axes labels. You can change the default formatting by selecting **X-Y Plot Format** under the **Graphics** menu. When you click on an existing graph, a selection box appears around the graph, and the name of the **Graphics** menu becomes X-Y Plot. (Note: This change in the name of the **Graphics** menu does not occur in Mathcad 4.0.) By selecting **Format** under the **X-Y Plot** menu, you can change one or both axes from linear to logarithmic, add labels to the X and Y axes, and perform other operations. Double clicking on any graph or plot immediately takes you to the formatting dialog box.

Try It

Graph the altitude of a projectile for $t = 0$ to $t = 20$ seconds given that its altitude is defined by

$$\text{position}(t) := 490\cdot m - \frac{9.8\cdot\frac{m}{sec^2}}{2}\cdot(t - 10\cdot sec)^2$$

Use the **Format** option under the **X-Y Plot** menu to add an appropriate label to the graph.

Graphing multiple functions is often a simple way to determine a solution to simultaneous equations. Using a graph, one can see, at least roughly, the point of intersection of two or more curves. Such analysis can be used, for example, in engineering economics, to determine the breakeven point in the production of a product, as shown in Example 4-2.

EXAMPLE 4-2

Graphing Multiple Functions

The cost of producing a product generally decreases with the number of units produced. Of course, revenue increases with the number of units sold. In engineering economics, the breakeven point is that number of units produced and sold when costs equal revenues. Assume that analysis has shown that the cost, in dollars, of producing w widgets is given by the equation $cost(w) = 1{,}300 + 5.60*w$. That is, it costs \$1,305.60 to produce one widget, but it costs \$6.90 per widget to produce 1,000 widgets. If we sell our widgets for \$8.00 each, our revenues are defined by the equation *revenue*(*w*) = 8*w.

Plot the two functions in order to determine the breakeven point, that is, where the expenses equal the income.

SOLUTION

Clear the document window.

Define the range of the independent variable *w*, the number of widgets produced.

Type: `widgets:0;1000`(ENTER)

This time, we will define the functions outside the graph. This approach keeps the graph more concise, especially if the functions are particularly complex.

Type: `costs(widgets):1300+5.60*widgets`(ENTER)

Type: `revenues(widgets):8*widgets`(ENTER)

Select **Create X-Y Plot** under the **Graphics** menu.

In the placeholder at the bottom center of the graph, type: `widgets`

Click on the placeholder to the left of the center of the Y-axis.

To plot multiple functions on one graph, you enter the functions, or their names, separated by commas. In this case, you enter their names.

Type: `costs(widgets),revenues(widgets)`(ENTER)

You should see

widgets := 0 .. 1000

costs(widgets) := 1300 + 5.60·widgets

revenues(widgets) := 8·widgets

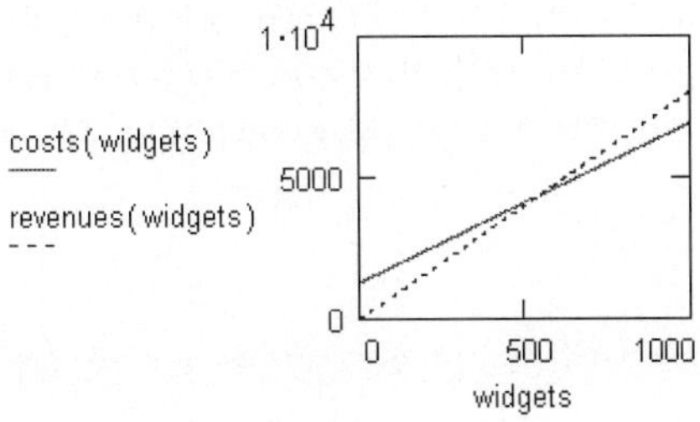

It can be seen from the graph that the breakeven point occurs approximately when 500 widgets are manufactured. The costs and revenues at that point are approximately \$4,000. Mathematical analysis would yield more accurate answers; however, an engineer often uses graphical analysis for quick, rough answers when accurate input figures, such as those for costs and revenues, are not available.

Try It Plot the following three functions over the range $x = -1$ to 0 in 0.01 increments:

- $f(x) = x^4 - x$
- $g(x) = x^2 - 1$
- $h(x) = -x + 1$

Graphing Multiple Vectors

As shown in Chapter 3, plotting the contents of a vector is similar to plotting a function. You define a range variable that is the index of the vector, create the vector if it does not already exist, and then create the graph. The index is used as the independent variable of the graph, the vector name as the dependent variable. It does not matter how the vector acquired its contents. The contents can be entered manually by you, they can be defined in a manner similar to defining a function, or they can be read from a data file.

EXAMPLE 4-3

Graphing Multiple Vectors

Plot the contents of a vector that contains the following numbers: -1.1, -3.2, -2.3, -1.1, 0.0, 1.0, 1.5, 1.1, 1.8, 2.1, and 3.0. On the same graph, plot a vector, the contents of which are defined by

$$X_i = \sin(2\pi i/50),\ i = 0..10$$

SOLUTION

Since the elements in the first vector cannot be defined by any easily discerned function, you must enter them manually. One approach would be to use the Matrices dialog box, which appears when you select **Matrices** under the **Math** menu, to create the vector, which we will call **Z**. You would define **Z** to be a matrix with 11 rows and 1 column. However, since **Z** is a vector, we will use an easier approach. First, since **Z** contains 11 elements, define an index that ranges from 0 to 10.

Type: `i:0;10`

Now create **Z**.

Type: `Z[i:-1.1,-3.2,-2.3,-1.1,0.0,1.0,1.5,1.1,1.8,2.1,3.0`(ENTER)

Notice that a column vector was created and shown in a box. The [told Mathcad you were defining a vector and i is its index. You separated the elements with commas.

To create the vector defined by $X_i = \sin(2\pi i/50)$, click to the right on the top of the **Z** vector and

Type: `X[i:sin(2*`(CTRL)`+p*i/50)`(ENTER)

Note that **X** uses the same index variable as does **Z**, that is, *i*.
Now create the graph.

Click to the right center of the **Z** vector and select **Create X-Y Plot** under the **Graphics** menu.

In the placeholder at the bottom center of the graph, type: `i`

Click on the placeholder to the left of the center of the Y-axis.

Type: `Z[i`(↑)(↑)`,X[i`(ENTER)

The screen should now show

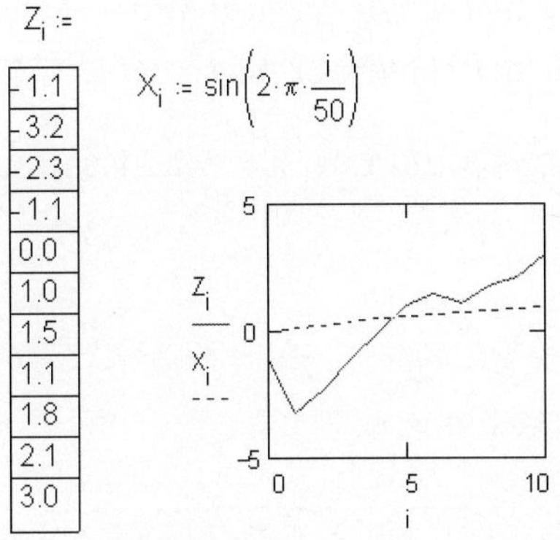

In the example, the data in each vector were plotted against their indices. Those indices might represent time in seconds or the number of widgets. It is sometimes desirable to plot the contents of one vector against the contents of another, as shown in Example 4-4.

EXAMPLE 4-4

Graphing One Vector Against Another

In electrical engineering, two or more different electronic signals are sometimes combined to produce certain desired effects. For example, two or more signals are used to drive the magnetic yokes in televisions and in radar displays in order to control the movement of the electron beam that draws images on the screen. An interesting pattern, called a Lissajous pattern, is produced when independent sine waves are applied to the X and Y inputs of an oscilloscope. To see what a Lissajous pattern looks like, plot the contents of a vector that is defined by

$$X_i = sin(2\pi i/50),\ i = 0..100$$

against the contents of a vector that is defined by

$$Y_i = cos(6\pi i/50),\ i = 0..100$$

SOLUTION

Begin by defining the index as a range variable.

Type: `i:0;100`

Next, define the contents of the two vectors.

Type: `X[i:sin(2*`(CTRL)`+p*i/50)`(ENTER)

Type: `Y[i:cos(6*`(CTRL)`+p*i/50)`(ENTER)

Now create the graph.

Select **Create X-Y Plot** under the **Graphics** menu.

In the placeholder at the bottom center of the graph, type: `X[i`

In the placeholder at the center left of the graph, type: `Y[i`(ENTER)

The screen should now show a Lissajous pattern.

$i := 0..100$

$X_i := \sin\left(2\cdot\pi\cdot\frac{i}{50}\right)$

$Y_i := \cos\left(6\cdot\pi\cdot\frac{i}{50}\right)$

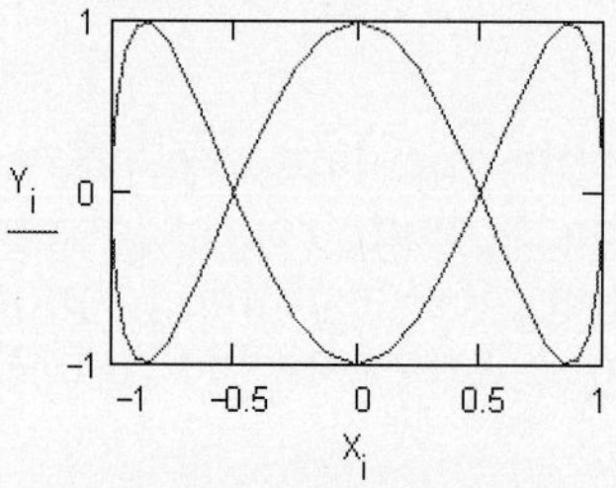

Try It

Plot the contents of a vector that is defined by

$X_i = sin(4i),\ i = 0..32$

against the contents of a vector that is defined by:

$Y_i = cos(4i),\ i = 0..32$

Graphing Histograms

In Section 3-3, you were shown how a histogram displays the frequency with which a number or set of numbers appears within a data set. Example 4-5 shows you how to plot a histogram in a graph.

EXAMPLE 4-5

Graphing a Histogram

Create a 1,000-element vector containing random real numbers that range from 100 to 200 kilograms. Then plot a histogram that shows how many of these random numbers lie between 100 and 110, 110 and 120, ..., and 190 and 200.

SOLUTION

Using a range variable *i*, create the 1,000-element vector **V**.

Type: `i:0;999`(ENTER)

Type: `V[i:(100+rnd(100))*kg`(ENTER)

Now define a range variable and use it to define the interval vector (see Section 3-3).

Type: `j:0;10`(ENTER)

Type: `boundaries[j:(100+10*j)*kg`(ENTER)

Note that the interval vector, called *boundaries*, has 11 elements because there are 10 intervals in which we are interested, requiring 11 boundaries to define them. The function *hist*(**int**,**v**) will then return a 10-element vector, indicating the contents of the 10 groupings of numbers.

If you wanted to display the contents of the histogram, you would type

`hist(boundaries,V)=`(ENTER)

However, here you want to plot the histogram. To do so, you create a function of the histogram and then graph that function in bar chart format.

First, create the function, calling it *f*.

Type: `f:hist(boundaries,V)`(ENTER)

Now, define the number of bars to be plotted (10).

Type: `k:0;9`(ENTER)

Now create the graph.

Select **Create X-Y Plot** under the **Graphics** menu.

In the placeholder at the bottom center of the graph, type: `k`

In the placeholder at the center left of the graph, type: `f[k`(ENTER)

You should now see a line graph. To convert it to a bar graph, click on the graph and then

select **Format...** under the **X-Y Plot** menu.

The **Formatting Currently Selected X-Y Plot** dialog box appears. It includes several rows and columns of formatting information.

Click on the row titled trace 1 if it is not already highlighted.

Click on the down arrow under the Type column.

Select *bar* (default is *lines*). Click on the OK button.

Your graph may be different from the one shown here because you are likely to generate a different set of random numbers. However, you should now see something like

$i := 0 .. 999$

$V_i := (100 + \mathrm{rnd}(100)) \cdot kg$

$j := 0 .. 10$

$boundaries_j := (100 + 10 \cdot j) \cdot kg$

$f := \mathrm{hist}(boundaries, V)$

$k := 0 .. 9$

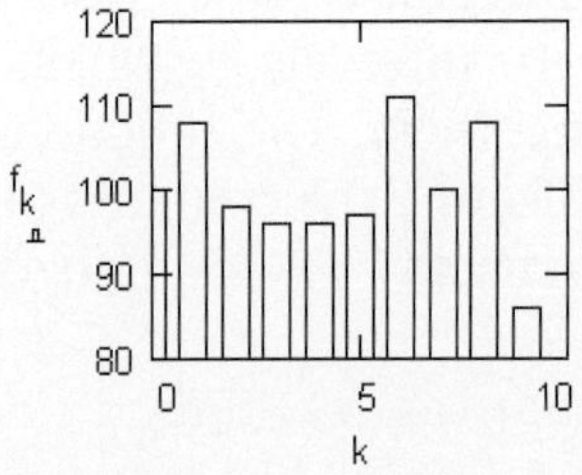

A histogram of the distribution of 100 random numbers ranging from 100 to 200.

The text to the right of the histogram was added to label the graph; you should add similar annotation to the plots you create.

Try It Create a 50-element vector containing random real numbers that range from 0 to 10. Then plot a histogram that shows how many of these random numbers lie between 0 and 1, 1 and 2,..., and 9 and 10.

Graphing Least Squares Fits

One of the most commonly performed operations on data sets is the *least squares fit*, also known as *linear regression.* Given a set of X-Y pairs of data, the goal is to plot a linear function that best describes data that reflect a linear, or somewhat linear, relationship. Mathcad includes two built-in functions required to determine this linear function. The function *slope*(**vx,vy**) returns the slope of a linear function that best fits the data pairs in vectors **vx** and **vy**. The function *intercept*(**vx, vy**) returns the Y-axis intercept of the same linear function. That is, given a set of X-Y data pairs, we require two vectors, one containing the X values, and the other containing the corresponding Y values. (Mathcad 6.0 Plus includes a *regress* function that makes the determination of the slope and intercept unnecessary. The *regress* function can also determine higher order data fits, which may better describe the data.)

When plotting a least squares fit curve, you first define a range variable as a vector index. You then create two vectors that, respectively, contain the known X and Y values. Then use the *slope* and *intercept* functions to compute the slope and intercept of the least squares fit curve. Lastly, plot the **y** vector and the least squares fit curve, both against the **x** vector. Example 4-6 illustrates the process.

EXAMPLE 4-6

Graphing a Least Squares Fit

Plot a least squares fit curve through the following data, which represent average daily oil production, in barrels, for a single oil well for each of 12 consecutive months: 70, 64, 65, 60, 68, 58, 54, 54, 58, 48, 49, 53.

SOLUTION

First, we will define the independent variable month, which is the index of the vector that contains the production data. Since the Mathcad functions we will be using require two vectors as arguments, we will define month as the contents of one vector rather than as a range variable. However, that vector itself mandates that we define a range variable *i*.

Type: `i:0;11`(ENTER)

Now, define the month vector to range from 1 to 12.

Type: `month[i:i+1`(ENTER)

Next, create the oil production vector.

Type: `prod[i:70,64,65,60,68,58,54,54,58,48,49,53`(ENTER)

Now determine the slope and Y-axis intercept of the curve that best fits this data.

Type: `a:slope(month,prod)`(ENTER)

Type: `b:intercept(month,prod)`(ENTER)

Finally, create the graph and plot the least squares fit function.

In order to place the graph to the right of the work done so far, click the mouse to the right of the *prod* vector, a bit below the center of the vector.

Select **Create X-Y Plot** under the **Graphics** menu.

In the placeholder at the bottom center of the graph, type: `month[i`

In the placeholder at the center left of the graph, type:
`prod[i`(↑)(↑)`,a*month[i`(↑)(↑)`+b`(ENTER)

The document window should now show

$i := 0..11$

$month_i := 1 + i$

$prod_i :=$

70
64
65
60
68
58
54
54
58
48
49
53

$a := \text{slope}(month, prod)$

$b := \text{intercept}(month, prod)$

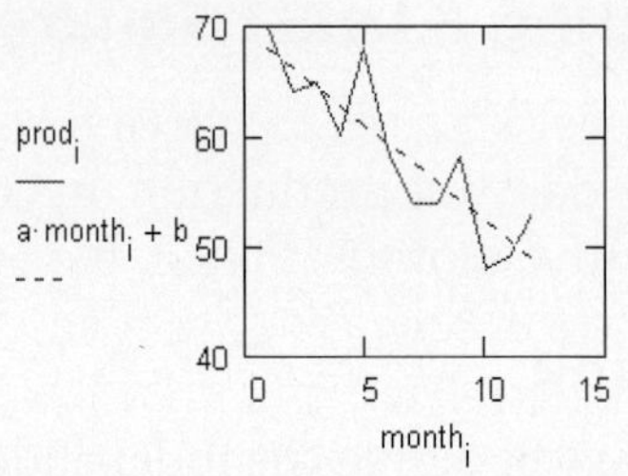

From a printout of this curve, one could easily extend the least squares fit line and estimate the month in which the oil production will reach zero. Of course, that month could also be determined mathematically from the equation for the least squares fit curve.

Try It Plot a least squares fit curve through the following data points: (1,1), (3,2), (4,4), (5,5), (7,7), and (9,5). Determine the point at which the least squares fit curve intersects the X-axis.

4-2 PLOTS

Plotting polar, surface, and contour plots is similar to plotting functions. The basic difference is the structure or format of the graph.

Creating Polar Plots

The graphs you plotted in Section 4-1 were all plotted using Cartesian coordinates, that is, on graphs using the familiar X and Y axes. However, many functions are not easily plotted using Cartesian coordinates or convey more information when plotted using coordinates other than Cartesian. For example, functions that involve trigonometry often convey more information when displayed using a *polar plot.* Antenna patterns and electrical or magnetic fields are often better described using polar plots.

A polar plot is begun by first defining a range variable, often ranging from 0 to 2π radians or from 0 to 360 degrees. The function is then defined. It can be defined directly as a function, or the function may be used to fill a vector with data points that are then plotted. Then you select **Create Polar Plot** under the **Graphics** menu and enter the independent and dependent variables in the plot's placeholders appropriately. Example 4-7 demonstrates the process.

EXAMPLE 4-7

Creating a Polar Plot of a Vector

Create a polar plot of the following two functions:

$$r(\theta) = 2^{0.25\theta} \text{ and } g(\theta) = 2^{-0.25\theta},\ \theta = 0..360 \text{ degrees}$$

SOLUTION

Start by defining a range variable.

Type: `i:0;360`(ENTER)

Now click to the right of the work you just entered and create a vector θ, of 360 angles, measured in radians.

Type: `q`(CTRL)+`g[i:i*2*`(CTRL)+`p/360`(↑)(↑)(↑)`*rad`(ENTER)

Typing (CTRL)+g changes the variable *q* to its Greek equivalent θ. You don't have to use Greek variables, but it is customary when dealing with trigonometric functions. The units of the *i* elements in θ are radians. Note that what you did here was convert *i*, which varies from 0 to 360 degrees, to θ_i, which contains elements that vary from 0 to 6.283 radians. (There are 2π, or 6.283, radians in 360 degrees.)

Now define the function *r*, in meters.

Type: `r[i:2^0.25*q`(CTRL)+`g[i`(↑)(↑)(↑)(↑)`*m`(ENTER)

Define the function *g*, in meters.

Type: `g[i:-2^-0.25*q`(CTRL)+`g[i`(↑)(↑)(↑)(↑)`*m`(ENTER)

To create the polar plot

select **Create Polar Plot** under the **Graphics** menu (or use the Graphing Palette on the palette).

Now plot the vector θ against the vectors **r** and **g**.

In the placeholder at the bottom of the plot, type: `q`(CTRL)+`g[i`

In the placeholder to the left of the plot, type: `r[i`(↑)(↑)`,g[i`(ENTER)

You should now see

$i := 0..360 \qquad \theta_i := i \cdot 2 \cdot \frac{\pi}{360} \cdot rad$

$r_i := 1.1^{0.25 \cdot \theta_i} \cdot m \quad g_i := -1.1^{0.25 \cdot \theta_i} \cdot m$

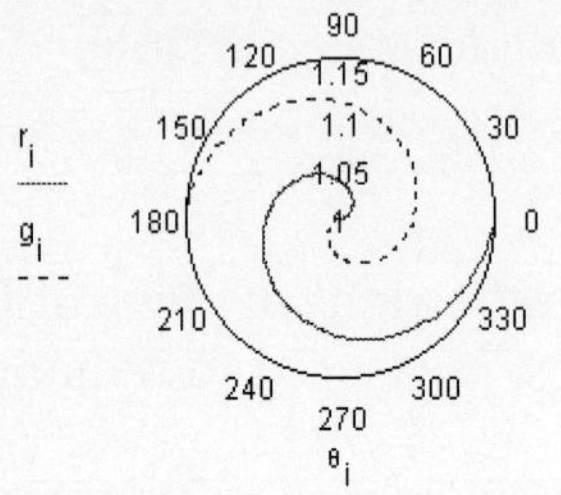

The numbers 0, 1, and 2, in the interior of the plot represent the lengths of the radii at those points. Note that at θ = 0, the values of both functions are 1 meter. As θ increases, *r*(θ) spirals outward while *g*(θ) spirals inward.

A more direct approach to Example 4-7 would have been to use the angle θ as the index of the two functions plotted. However, θ must be an integer because it is the index of a vector. If θ had been used in the creation of the vectors, rather than θ_i, a less accurate plot would result. To overcome this problem, in Example 4-8, which repeats the previous example, the graphs are plotted directly as functions, in which θ need not be an integer.

EXAMPLE 4-8

Creating a Polar Plot of a Function

Create a polar plot of the two functions:

$$r(\theta) = 1.1^{0.25\theta} \text{ and } g(\theta) = -1.1^{0.25\theta},\ \theta = 0..360 \text{ degrees}$$

SOLUTION

Start by defining the range variable, incrementing it in 0.1 increments, and giving it the units degrees.

Type: `q(CTRL)+g:0*deg,0.1*deg;360*deg(ENTER)`

Now define the function *r*, in meters.

Type: `f(q(CTRL)+g):2^0.25*q(CTRL)+g↑↑↑*m(ENTER)`

Now define the function *g*, in meters.

Type: `g(q(CTRL)+g):2^-0.25*q(CTRL)+g↑↑↑↑*m(ENTER)`

To create the polar plot

select **Create Polar Plot** under the **Graphics** menu.

Now plot the vector θ against the vectors **f** and **g**.

In the placeholder at the bottom of the plot, type: `q`(CTRL)`+g`

In the placeholder to the left of the plot, type:
`f(q`(CTRL)`+g),g(q`(CTRL)`+g)`(ENTER)

You should now see

$f(\theta) := 1.1^{0.25\theta} \cdot m \quad g(\theta) := -1.1^{0.25\theta} \cdot m$

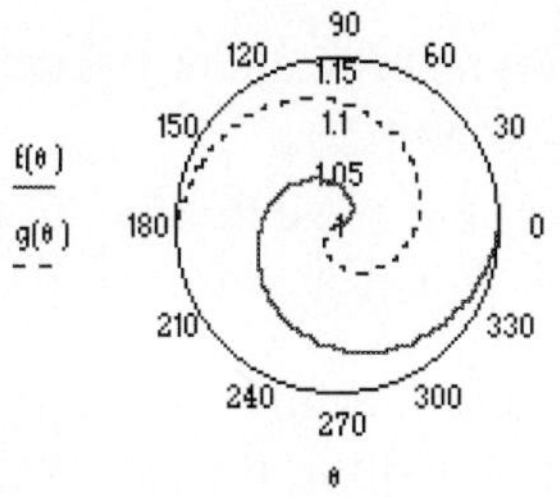

When you click on a polar plot, the name of the **Graphics** menu changes to **Polar Plot**. (Note: This change in the name of the **Graphics** menu does not occur in Mathcad 4.0.) Select **Format** under the **Polar Plot** menu (or simply double click on the plot) to make formatting changes, such as changing to a logarithmic scale or adding grid lines.

Try It Create a polar plot of the functions

$$f(\theta) = \sin(\theta) + 1 \text{ and } g(\theta) = -\sin(\theta) + 1,\ \theta = 0..360 \text{ degrees}$$

in 1-degree increments.

Creating Surface Plots

A *surface plot* often is used to plot the surface of three-dimensional functions or a land surface. To create a surface plot, you must first have a matrix, or two-dimensional array, of values to plot. Mathcad uses the row and column numbers of the matrix as the distances along the X- and Y-axes. The elements of the matrix are plotted as heights above or below the X-Y plane, that is, along the Z-axis. The values of the matrix elements can be defined in three ways: They can be entered as specific numbers, or they can be defined by a two-dimensional function, or they can be read from a data file using the built-in *READPRN*(file) function.

EXAMPLE 4-9

Creating a Surface Plot

Create a surface plot of the following function: $f(x,y) = \sin(x^2 + y^2)$ for $x = -2$ to $+2$ and $y = -2$ to $+2$.

SOLUTION

We must first decide at what resolution we want to create the plot. Let's plot 16 points by 16 points. Therefore, we will need a 16×16 matrix. To create the two indices using range variables, type: `i:0;15`(ENTER)

Click to the right of the definition of *i* and type: `j:0;15`(ENTER)

Now create two vectors that will hold the values of *x* and *y*, which will act as the independent variables of the two axes. The trick is to determine the function that will give you 16 values of *x* and *y* evenly spread between -2 and +2, as specified in the problem. This is exactly what the following does:

Click at the left margin, under the definition of *i*, and type:
`x[i:-2+i/3.75`(ENTER)

Click to the right of the definition of the vector *x*, and type:
`y[j:-2+j/3.75`(ENTER)

Now click at the left margin, under the definition of **x**, and define the function to be plotted, calling it **f**. Then create the matrix, calling it **M**:

Type: `f(x,y):sin(x^2`(↑)(↑)`+y^2)`(ENTER)

Click to the right of the definition of *f(x,y)* and type:
`M[i,j:f(x[i`(↑)(↑)`,y[j)`(ENTER)

Now plot the function. Click at the left margin, under the definition of *f(x,y)* and then

Select **Create Surface Plot** under the **Graphics** menu.

In the placeholder at the bottom of the plot, type: `M,`(ENTER)

You should now see

$i := 0 .. 15 \qquad j := 0 .. 15$

$x_i := -2 + \dfrac{i}{3.75} \quad y_j := -2 + \dfrac{j}{3.75}$

$f(x,y) := \sin\left(x^2 + y^2\right) \qquad M_{i,j} := f\left(x_i, y_j\right)$

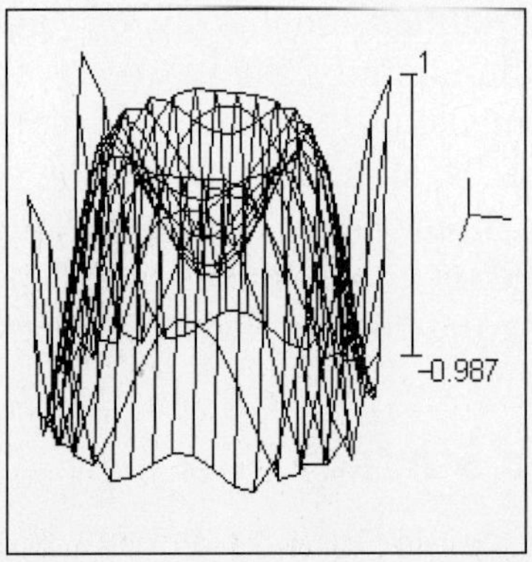

M

You can format surface plots in a number of ways. To see the available options, select **3D Plot Format** under the **Graphics** menu or double click on the plot. Surface plots can be titled, resized, rotated, tilted, and shaded. They can also be converted to contour plots or 3D bar charts.

Try It Create a surface plot of the function $f(x,y) = \tanh(x^2 + y^2)$ for $x = -1$ to $+1$ and $y = -1$ to $+1$. Format the plot, tilting it 90 degrees and selecting Hide Mesh as its Fill Style (see the Color & Lines option). Convert the plot into a 3D bar chart.

Creating Contour Plots

The only difference between a surface plot and a *contour plot* is that a surface plot is displayed as a three-dimensional surface, whereas a contour plot is displayed as a two-dimensional surface, with lines of contour shown.

You create a contour plot in much the same way as you create a surface plot. Either way, you plot a 2D matrix. In the case of a contour plot, you might be plotting, not a mathematical function, but points of elevation on a piece of land. In this case, you would have to enter the elevation values manually into either a matrix created in a Mathcad document or into a data file that is then read into a matrix in a Mathcad document. As we explained in Section 3-2, you can use any editor or word processor to enter data into a data file. When creating a contour plot, the data would be entered in rows, just as they would appear in the matrix, with the values separated by spaces and the rows separated by carriage returns. The file must be saved as an unformatted ASCII character file.

You can easily change a surface plot into a contour plot; they are just two different ways to graphically visualize the same data. To do so, double click on the plot. Notice in the 3D Plot Format dialog box that appears that you can convert any 3D plot (surface, contour, or 3D bar chart) into any other form of 3D plot. (Mathcad versions 4.0 and 5.0 vary considerably in how they format and convert plots.)

EXAMPLE 4-10

Creating a Contour Plot

Create a contour plot of the elevations, in feet, of a plot of land. The data are stored in a 10×10 matrix in a file called elevate.prn. The file was created by using the Write program in the Windows Accessories group. Remember that Mathcad uses *READ* to read single values, normally for vector input, from files that have a .dat extension. Mathcad uses *READPRN* to read entire two-dimensional data files into matrices; *READPRN* assumes the file has the extension .prn. Be sure any data files you want Mathcad to read are simple text files.

SOLUTION

Start by reading the contents of the data file into a matrix **X**. Remember that, although the data file name (elevate.prn) must have a .prn extension, that extension is not used when specifying the file name in the *READPRN* command.

Type: `X:READPRN(elevate)`(ENTER)

To display the contents of elevate.prn, type: X=(ENTER)

Select **Create Contour Plot** under the **Graphics** menu.

In the placeholder at the bottom of the contour plot box, type: X

You should now see the contents and the contour plot of elevate.prn:

X := READPRN(elevate)

X =

100	100	100	100	100	100	100	100	100	100
100	110	110	110	110	110	110	100	100	100
110	100	120	120	120	120	110	110	100	100
100	110	120	130	140	120	120	120	110	100
100	110	120	130	140	130	130	120	110	100
100	110	120	120	130	130	130	120	110	100
100	110	120	120	120	120	130	120	110	100
100	100	110	110	110	120	120	110	110	100
100	100	100	100	110	110	110	100	100	100
100	100	100	100	100	100	100	100	100	100

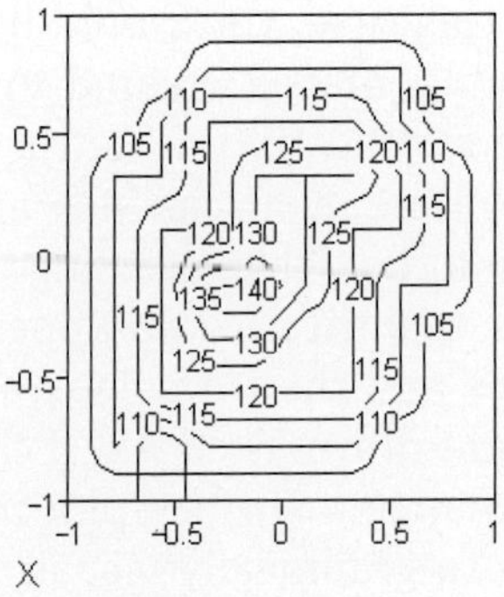

The axis limits on the contour plot have no significance as coordinates. They affect the display only and can be changed manually.

A contour plot can be converted to a surface plot in the same way a surface plot is converted to a contour plot. Double click on the plot and then select Surface Plot under the View option. Grayscale can be added by selecting it under the Color & Lines options.

Try It Create a contour plot of the function $f(x,y) = \tanh(x^2+y^2)$ for $x = -1$ to $+1$ and $y = -1$ to $+1$. Format the plot, adding grayscale. Then convert the plot to a surface plot. Finally, convert the plot to a 3D bar chart.

Creating Scatter Plots

Scatter plots display any given collection of data points in three-dimensional space. Scatter plots allow you to identify clusters of data or to plot the trajectory of an object through space. Scatter plots differ from Mathcad's other 3D plots in that they all allow you to plot several z values for the same *x* and *y* values. This is accomplished by using a different way to present the data: While in other 3D plots, the data are stored in 2D matrices, the data for a scatter plot are stored in three vectors. Each vector has as many elements as there are points to plot, and the *x*, *y*, and *z* coordinates of a point go into the three elements of the corresponding vectors. Example 4-11 shows how it works.

EXAMPLE 4-11

Creating a Scatter Plot

Create a scatter plot of the following three sets of temperature data measured over the surface of a 2 cm × 2 cm square metal plate at three different times.

Temperatures (C°) at Time 1.

2cm	88	88	87
1cm	85	85	84
0cm	83	82	83
y/x	0cm	1cm	2cm

Temperatures (C°) at Time 2.

2cm	75	75	74
1cm	74	74	73
0cm	72	71	71
y/x	0cm	1cm	2cm

Temperatures (C°) at Time 3.

2cm	62	60	60
1cm	60	59	59
0cm	58	58	57
y/x	0cm	1cm	2cm

SOLUTION

To create a scatter plot, you start by creating three vectors, each having as many elements as you have points to plot. We have 3 × 9 = 27 points to plot. Start by creating the index for the three vectors.

Type: `i:0;26` (ENTER)

Then create the vectors.

Type:
`x[i:0,1,2,0,1,2,0,1,2,0,1,2,0,1,2,0,1,2,0,1,2,0,1,2,0,1,2`
(ENTER)

Type:
`y[i:2,2,2,1,1,1,0,0,0,2,2,2,1,1,1,0,0,0,2,2,2,1,1,1,0,0,0` (ENTER)

Type:
`temp[i:88,88,87,85,85,84,83,82,83,75,75,74,74,74,73,72,71,71,62,60,60,60,59,59,58,58,57` (ENTER)

Note that the temperatures were entered by row, starting at the top row of the first set of temperatures and working down to the bottom row of the third set of temperatures.

Now select **Create 3D Scatter Plot** under the **Graphics** menu.

In the placeholder at the bottom of the scatter plot box, type: `x,y,temp`

Double click on the plot and format it by adding color (see Shading), grid lines on all three axes (see Axes), and perimeter axes (see View).

You should now see the following display (the bottom has been truncated here due to space limitations):

$i := 0 .. 26$

$x_i :=$	$y_i :=$	$temp_i :=$
0	2	88
1	2	88
2	2	87
0	1	85
1	1	85
2	1	84
0	0	83
1	0	82
2	0	83
0	2	75
1	2	75
2	2	74
0	1	74
1	1	74
2	1	73
0	0	72
1	0	71
2	0	71
0	2	62

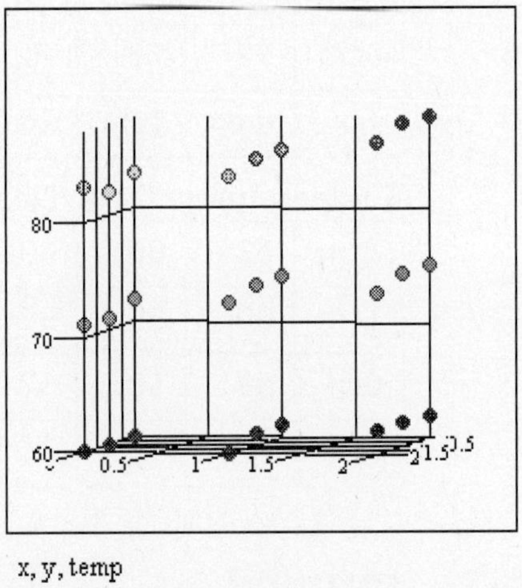

Note that you can perform a great deal of customization of a scatter plot by double clicking on to bring up the formatting options.

Try It The trajectory of an object through 3D space is described by the function $z(x,y) = \tanh(x^2+y^2)$. Create a scatter plot of this trajectory, using three 25-element vectors. The x and y vectors should contain equally distant values that range approximately from −1 to +1; the **z** vector contains the appropriate corresponding z values. (Note that in Example 4-11, you entered the vector values manually; in this problem, you will want to use functions to create all three vectors.)

Application 3 TOPOGRAPHIC SURVEYING

Civil engineers design and construct major structures and facilities such as bridges, dams, tunnels, buildings, factories, highways, airports, and railroads. A fundamental activity in the construction industry is surveying. Let's assume we need a topographical survey of a square 60 acres of hilly land that must be leveled for construction of a shopping center. The survey will allow engineers to determine how much earth must be moved or removed to level the land.

1. Define the Problem

Surveying measurements are made vertically, on a plumb line directed toward the earth's center, and horizontally measured perpendicular to a plumb line. All objects bear relationships to each other that can be expressed by the horizontal direction, horizontal distance, and vertical distance between them. Surveying is a technique to measure and lay out these three dimensions. The usual procedure is to set a framework of convenient points and accurately determine their relationships—horizontal distances and directions and elevations.

When a construction project involves the movement of earth, the civil engineer must determine the altitude of enough points on the land so that he or she can accurately determine how much earth must be moved or removed in order to level the land. The engineer must first determine how many points must be measured. Surveying too few will not give an accurate measure; surveying too many will be unnecessarily expensive.

2. Gather Information

In this case, we have 60 acres of land; that is, 2,793,600 square feet. If the engineer decides to measure altitudes every 100 feet in each direction, he or she must make 256 measurements. These measurements must be made starting at one corner of the square piece of land, every 100 feet in each direction, creating 16×16 matrices of numbers. The data will be stored in a 16×16 data file.

3. Generate and Evaluate Potential Solutions

The land can be surveyed in several ways. The plumb line method described in step 2 is one approach. There are other available methods, including the use of lasers for accurate measurements. However, in this case, the less expensive plumb line method is adequate since the measurements need not be extremely accurate.

4. Refine and Implement a Solution

The data are stored, using an on-site laptop computer, in a file called survey.prn. Survey.prn is an ASCII character file, consisting of 16 lines of 16 data values, measured in feet, as follows:

```
100 103 103 103 102 101 101 100 100 103 102 101 101 102 102 100
101 105 105 103 102 101 104 102 101 103 102 101 108 108 106 100
101 105 110 105 103 101 104 103 101 103 102 101 110 110 109 100
102 103 105 103 101 101 104 104 101 103 102 108 112 115 112 100
100 102 102 100 100 101 104 104 102 103 102 106 110 112 109 100
103 105 100 103 98 98 104 104 103 103 102 101 105 108 107 100
108 105 102 100 96 94 90 95 98 103 102 101 102 105 105 100
104 103 101 103 97 96 93 94 97 103 102 101 101 104 103 100
102 102 100 102 100 98 99 100 102 104 102 101 101 100 102 100
100 101 100 101 102 101 104 103 102 101 102 101 96 96 98 100
103 101 100 100 100 101 104 104 101 101 99 96 91 88 88 96
102 101 100 101 99 101 104 102 102 102 100 94 90 82 86 92
101 102 101 101 99 101 104 102 102 101 100 96 93 80 84 90
102 103 100 100 100 101 103 102 102 102 101 98 98 86 88 100
101 102 101 101 102 101 102 102 102 101 101 101 100 94 99 99
100 101 100 100 102 101 101 101 102 101 102 101 101 100 100 100
```

The following shows how simple it is to use Mathcad to read and plot the data:

S := READPRN (survey) · ft

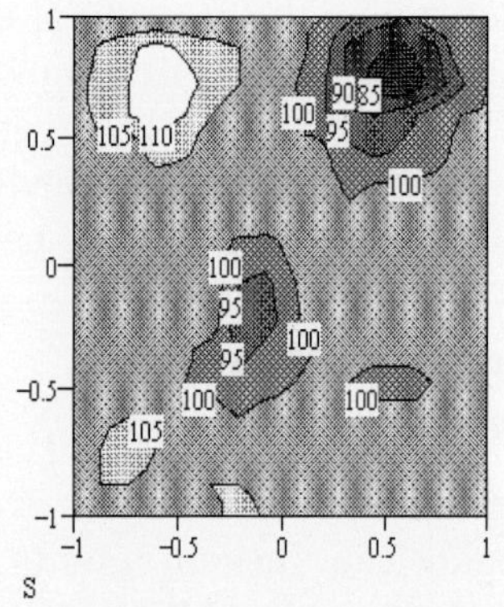

max(S) = 115 · ft   min(S) = 80 · ft

Using the **Format** selection under the **Contour Plot** menu, the plot was shaded. The light shading indicates areas of higher elevation, and the darker shading indicates areas of lower elevation. From the plot, the engineer can see the areas from which, and to which, earth will have to be moved.

5. Verify and Test the Solution

One could compute manually the amount of earth that must be moved or removed once the measurements are made. Alternatively, the data matrix could be fed into a program that then performs the computations. There is no way to test the accuracy of the engineer's work other than to actually perform the work and then measure the amount of earth that was moved or removed.

What If

What if a geological survey shows that there is rock just beneath the surface of much of the land in question? What might be the alternatives to making perfectly level the land on which the shopping center will be constructed? What factors might affect the cost of construction?

SUMMARY

This chapter showed you how to create graphs and plots. We began with an explanation of the graphing of functions and vectors. We next covered the creation of histograms and then discussed the computation and graphing of a least squares fit of data. Descriptions of the drawing of polar, surface, contour, and scatter plots followed. Throughout, it was emphasized how easily you can format Mathcad plots to meet your needs. Simply double clicking on any plot brings up the formatting window, where you can specify, among other things, logarithmic axes, labels, and scaling. The chapter concluded with a civil engineering application.

Key Words

Cartesian
contour plot
histogram
least squares fit
polar plot
linear regression
scatter plot
surface plot
X-Y plot

Problems

Note: All graphs and plots should be printed.

1. Graph the function $y = -(3/4)x + 5$ for $x = -10$ to $+10$.
2. Graph the function $y(t) = 490 - (a/2)(t{-}10)^2$, where a equals 9.8 m/sec^2, the acceleration due to gravity, t is in seconds, and $y(t)$ is the altitude in meters. Determine the time t' at which the altitude is zero. In the graph, let t vary from 0 to t'.
3. On the same graph, plot the following functions over the range $x = -100$ to $+100$ in increments of 5:

   $f(x) = x^3 - (1/2)x$

   $g(x) = -50x^2$
4. Plot function $f(x) = -20(-x)^{1/2}$ over the range $x = 1$ to 2 in 0.001 increments.
5. Plot the following three functions over the range $x = 1$ to 2 in 0.01 increments:

   $f(x) = -20(-x)^{1/2}$

   $g(x) = x^{-2} - 10$

   $h(x) = -x^{-1} + 1$
6. Plot the contents of a vector that contains the numbers -1.2, -4.2, -3.3, -2.1, 1.0, 1.0, 1.5, 2.1, 3.8, 4.1, and 5.0. On the same graph, plot a vector, the contents of which are defined by

   $\mathbf{X}_i = -\cos(2\pi i/40),\ i = 0..10$

7. Plot the contents of a vector that is defined by

   $\mathbf{X}_i = \sin(2\pi i/50)$, $i = 0..100$

   against the contents of a vector that is defined by

   $\mathbf{X}_i = \cos(6\pi j/50)$, $i = 0..100$

8. Using Cartesian coordinates, plot the contents of a vector that is defined by

   $\mathbf{Y}_i = \sin(i)$, $i = 0..6$

   against the contents of a vector that is defined by:

   $\mathbf{Y}_i = \cos(i)$, $i = 0..6$

9. Create a 100-element vector containing random real numbers that range from 0 to 600 meters. Then plot a histogram that shows how many of these random numbers lie between 0 and 100, 100 and 200, ..., and 500 and 600 meters.

10. Create a 200-element vector containing random real numbers that range from -10 to 10. Then plot a histogram that shows how many of these random numbers lie between -10 and -5, -5 and 0, 0 and +5, and +5 and +10.

11. Plot a least squares fit curve through the following data that represent the pressure, in psi, in a tank, measured once daily for 21 days: 12, 14, 16, 17, 20, 20, 20, 25, 25, 30, 31, 35, 37, 40, 41, 44, 44, 48, 54, 54, 55.

12. Plot a least squares fit curve through the following data points: (0,5), (1,3), (4,4), (5,7), (7,8), (9,7), (11,10), and (12,12).

13. Create a polar plot of the functions

    $r(\theta) = 1.1^{-0.25\theta}$ and $g(\theta) = 1.1^{0.25\theta}$, $\theta = 0..360$ degrees.

14. Create a polar plot of the functions

    $f(\theta) = \sinh(\theta^{-1/2})$ and $g(\theta) = -\cosh(\theta^{-1/2})+1$, $\theta = 1..15$ degrees

    in 1-degree increments.

15. Assume you are a civil engineer who must make a rough estimate of how many cubic yards of earth must be removed in order to level a plot of land to 5 feet above sea level. The following data represent elevations measured every yard in each direction on the plot, the size of which is 7 by 7 yards.

5	5	5	5	5	5	5	5
5	6	6	6	6	6	6	5
5	6	7	7	7	7	6	5
6	7	7	8	8	7	7	6
6	7	7	8	8	7	7	6
5	5	5	5	5	5	5	5
5	6	6	6	6	6	6	5
5	6	7	7	7	7	6	5

    Store the 8×8 matrix in an ASCII character file called matrix.prn. Read the matrix and create a contour plot of its contents. Add grayscaling. Make a rough estimate of the amount of earth that must be moved through visual examination of the contour plot. Convert the plot into a surface plot.

16. Create an 11×11 matrix **M** in which the elements are defined in terms of the matrix indices in the following way: $M(i,j) = (i-5)(j-5)$, where i and j range from 0 to 10. Create a 3D bar chart of the matrix contents.
17. Determine the low and high points in a surface defined by $f(x,y) = e^{-x}e^{-y}$ for $x = 0$ to 10 and $y = 0$ to 10. Use a surface plot and note the minimum and maximum indicated on the Z-axis.
18. Determine the low and high points of the surface defined by $f(x,y) = \sin(x^2+y^2)-\cos(x^2+y^2)$ for $x = 0$ to 4 and $y = 0$ to 4. Use a surface plot and note the minimum and maximum indicated on the Z-axis. Compare a contour plot of the function.
19. Graphically determine the points between $x = 0$ and $x = 4$ where the $y(x)$ is zero when $y(x) = e^{-x}\sin(2x)$.
20. Assume a manufacturing process in which the cost overhead is $50,000. It costs $18.50 to manufacture each part. Graphically determine the economic breakeven point if the parts are sold for $30.00 each.
21. Create a surface of plot half of a unit sphere.
22. Create a surface plot of a unit sphere. (See the QuickSheets.)
23. The trajectory of an object through 3D space is described by the function $z(x,y) = -(x^2+y^2)$ where x and y both range from -100 to $+100$. Create a scatter plot of the trajectory, using three 50-element vectors.

5 Numerical Methods

Prime Movers

Mechanical engineering is concerned primarily with ways of converting energy to useful mechanical forms and in designing the resulting machines for efficiency. These machines consist of two major subsystems: prime movers and functional mechanisms. Prime movers convert energy into mechanical form; functional mechanisms convert the output of prime movers into useful work. The internal combustion engine exemplifies a prime mover; it converts the energy stored in gasoline into the movement of a rotating shaft. The transmission, drive shaft, differential, axle, and wheels of an automobile act as functional mechanisms, converting the output of the engine into movement of the vehicle. Whatever the prime mover or functional mechanism, one of the main issues for mechanical engineers is efficiency. That's why composite materials often are used in the construction of functional mechanisms. A composite consists of materials in which particles or fibers are dispersed or woven into some form of matrix. The materials are selected and manufactured to provide unique properties. One familiar composite is fiberglass; another is carbon fiber. The application in this chapter examines the construction of carbon fiber racing wheels.

INTRODUCTION

In Section 5-1 of this chapter, we will show you how to determine the derivative of a function at a particular point and how to use the Mathcad symbolic processor to determine the function that is the derivative of a given function. In Section 5-2 we will show you how to determine the integral of a function over an interval and how to use the Mathcad symbolic processor to determine the function that is the integral of a given function.

5-1 DIFFERENTIATION

Engineers and their inventions commonly measure such quantities as linear position, angular position, stress, strain, speed, acceleration, and temperature. They also often measure the rate of change of these quantities, as in miles per hour, revolutions per minute, and meters per second squared. To determine the rate of change of a function, one takes the *derivative* of that function. The act of taking the derivative is called *differentiation.*

For example, when one graphs a quantity against time, the derivative of the resulting curve at any time t is the rate of change of the quantity at time t. If a graph plots position versus time, the derivative of the curve represents speed; that is, if one has a mathematical function that describes the position, the derivative of that function describes the speed.

Evaluating Derivatives

In general, the derivative of any function represents the rate of change of that function with respect to the independent variable. That independent variable does not have to be time. It need not have units at all. Given any variable y that is a function of a variable x, the derivative of y with respect to x yields the rate of change of y with respect to x. That is, given a change in x, the derivative tells you how much y changes.

The second derivative represents the rate of change of the rate of change for the independent variable. For example, taking the derivative of a function that defines a distance would yield the rate of change of distance, or speed. The second derivative would yield the rate of change of speed, or acceleration. The third derivative would yield the rate of change of acceleration.

Mathcad gives you the tools to easily determine the derivative of a function and evaluate that derivative at a given point. Example 5-1 takes you through the few simple steps necessary to determine the derivative of a function at a particular point.

EXAMPLE 5-1

Determining a Derivative at a Point

Determine the derivative of the following functions at the point $x = 1$:

$$y(x) = 4x^2 - 5$$

$$y(x) = -3x^3 + x$$

SOLUTION

Click at the location on the screen where you want your work to appear.

Give Mathcad the independent variable value at which you want the derivative evaluated.

Type: `x:1`[ENTER]

Click on the Calculus Palette button on the palette. (The Calculus Palette button has an integral symbol on it. If you leave the mouse arrow momentarily over any button on the palette, Mathcad will display a concise description of the button.) Then select the derivative operator (d/dx) on the palette.

You should now see

$$x := 1$$

$$\frac{d}{d\blacksquare}\blacksquare$$

The placeholder on the right should have a selection box around it. Mathcad is waiting for you to enter the function to be differentiated. The placeholder in the denominator requires the name of the independent variable, *x* in this case.

Without pressing [ENTER], type: `(4*x^2`[↑][↑]`-5)`

(Note: Without the parentheses, only the $4x^2$ would be differentiated.)

Without pressing [ENTER], click on the placeholder in the denominator and type: `x=`

The derivative of $4x^2-5$ at $x = 1$ should appear to the right. (Make sure the Automatic Mode button on the tool bar is active so that calculations are performed as soon as the equals sign is pressed.) Optionally, you can enter units in the placeholder to the right of the result. Or, press [ENTER] and the placeholder will disappear. You should now see

$$x := 1$$

$$\frac{d}{dx}\left(4 \cdot x^2 - 5\right) = 8$$

We will determine the second derivative a bit differently, just to illustrate your options when using Mathcad.

Type: `y(x):-3*x^3`[↑][↑]`+x`[ENTER]

Select the derivative operator on the Calculus Palette.

In the right placeholder, without typing [ENTER], type: `y(x)`

Click on the placeholder in the denominator and type: x=(ENTER)

You should see

$$x := 1$$

$$\frac{d}{dx}\left(4 \cdot x^2 - 5\right) = 8$$

$$y(x) := -3 \cdot x^3 + x$$

$$\frac{d}{dx}y(x) = -8$$

Try It Determine the derivative of $\sin(x)^2$ at $x = \pi/4$.

Symbolic Differentiation

Sometimes you do not want the value of a derivative at a particular point; you just want the derivative function. Mathcad includes a symbolic processor for just this purpose. As we noted earlier, *symbolic processing* involves evaluating a function in any of a number of ways, including differentiation and integration, without substituting any values for the variables and without determining any numerical results. For example, when directed to determine the derivative of $3x^2$ at $x = 1$, Mathcad will return the result 6, that is, $6x$ evaluated at $x = 1$. The symbolic processor does not determine specific values. Asked to determine the derivative of $3x^2$, it will return the result $6x$.

The Mathcad symbolic processor is capable of a number of powerful symbolic operations. To see the available operations, select the **Symbolic** menu. Those options not presently available will not be highlighted. Three of the more useful commands are **Simplify**, **Expand Expression**, and **Factor Expression**. The **Simplify** command carries out basic simplification of algebraic expressions and applies trigonometric and inverse function identities. You use the **Expand Expression** command to multiply out powers and products of polynomials. Use this command also to write a rational expression with a polynomial numerator as a sum of fractions and to write an expression like sin(5x) as a polynomial in sine and cosine. The **Factor Expression** command tries to write the selected expression as a product. It converts polynomials into a product of polynomials, integers into a product of primes, and sums of rational expressions into single fractions.

Here, we examine only Mathcad's ability to perform symbolic differentiation and integration. In Example 5-2, you symbolically determine the derivative of the functions with which you dealt in the previous example.

EXAMPLE 5-2

Determining a Derivative Polynomial Function

Determine the first, second, and third derivatives of $y(x) = -3x^3 + x$

SOLUTION

Type: `y(x):-3*x^3`(↑)(↑)`+x`(ENTER)

Click on either *x* in the right side of the equation so that a blue insertion point appears to the left or to the right of the *x*.

Select **Differentiate on Variable** under the **Symbolic** menu.

The first derivative of *y* with respect to *x* should appear.

Click on the *x* in the right side of the derivative and again select **Differentiate on Variable** under the **Symbolic** menu.

Click on the *x* in the right side of the second derivative and again select **Differentiate on Variable** under the **Symbolic** menu.

You should now see

$$y(x) := -3\cdot x^3 + x$$

$$y(x) := -9\cdot x^2 + 1$$

$$y(x) := -18\cdot x$$

$$y(x) := -18$$

Mathcad can determine the derivative of trigonometric functions as well as polynomials, as shown in Example 5-3.

EXAMPLE 5-3 Determining a Derivative Polynomial Function

Determine the derivative of $y(x) = \cos(-3x)^2 - \sin(-3x)^2$.

SOLUTION

Type: `y(x):cos(-3*x)^2 - sin(-3*x)^2`(ENTER)

Click on either *x* in the right side of the equation so that a blue insertion point appears to the left or to the right of the *x*.

Select **Differentiate on Variable** under the **Symbolic** menu.

The first derivative of *y* with respect to *x* should appear, and the screen should show

$$(x) := \cos(-3\cdot x)^2 - \sin(-3\cdot x)^2$$

$$(x) := -12\cdot \cos(3\cdot x)\cdot \sin(3\cdot x)$$

Note that a function can involve a number of variables, and you can perform partial differentiation with respect to any variable in the function.

Try It Differentiate 490 - (9.8/2)(t - 10)^2 with respect to t.

5-2 INTEGRATION

The effect of *integration* is the opposite of that of differentiation. For example, given a function of speed versus time, the derivative yields acceleration versus time. The *integral* yields distance versus time. In fact, integration is the process of determining the area under the curve of a graph function. For example, the area under a speed curve represents distance traveled.

Evaluating Integrals

Solving for integrals using Mathcad is similar to solving for derivatives. In Example 5-4, you determine the area under a curve between two points on the curve.

EXAMPLE 5-4

Determining an Integral over an Interval

Determine the distance, in meters, a free-falling object falls in 5 seconds if it starts at rest. The velocity of the object $v(t)$ = gt, where $v(t)$ is in meters per second, g is the acceleration due to gravity in meters/second^2, and t is in seconds.

SOLUTION

Define the acceleration by typing: `g:9.86*m/sec^2`(ENTER)

Then define the time of free-fall by typing: `t:5*sec`(ENTER)

Define the velocity function by typing: `v(t):g*t`(ENTER)

Type: `"Speed of the object after 5 seconds`

Determine the object's speed after 5 seconds by typing: `v(t)=`(ENTER)

Type: `"Distance the object falls in first 5 seconds:`

Click below this text; then click on the Calculus Palette button. Select the integral operator that has limits (*a* and *b*).

In the placeholder to the left of the d, type: `v(t)`

In the placeholder to the right of the d, type: `t`

In the placeholder at the bottom of the integral symbol, type: `0*sec`

In the placeholder at the top of the integral symbol, type: `5*sec=`(ENTER)

You should now see

$$g := 9.86 \cdot \frac{m}{sec^2}$$

$$t := 5 \cdot sec$$

$$v(t) := g \cdot t$$

Speed of the object after 5 seconds

$$v(t) = 49.3 \cdot m \cdot sec^{-1}$$

Distance the object falls in the first 5 seconds

$$\int_{0 \cdot sec}^{5 \cdot sec} v(t)\, dt = 123.25 \cdot m$$

Although the distance the object falls in the first 5 seconds is 122.5 meters, the solution indicates the units as meters/second because those are the units of the function.

Note that the integral limits need not be constants. You can, for example, set the upper limit to, say, *i*, and then create a table of integral values by defining *i* as a range variable and the function as a vector.

Try It Determine the integral of $\sin(x)^2$ over the interval $x = 0$ to $x = \pi/4$.

Symbolic Integration

Sometimes you do not want the value of an integral over an interval, but rather desire the integral function. The Mathcad symbolic processor performs integration as well as differentiation. In Example 5-5, you symbolically determine the integral of two complex functions.

EXAMPLE 5-5

Determining an Integral Function

Integrate the following equation with respect to *x*:

$$y(t) = e^{-1}(sin(\pi t/8) + cos\ \pi t/8))$$

SOLUTION

Type: `y(t):e^-1`↑↑`*(sin(`CTRL`+p*t/8)+cos(`CTRL`+p*t/8))`ENTER

Click on either *t* in the right side of the function so that a blue insertion point appears on the left or right of the *t*.

Select **Integrate on Variable** under the **Symbolic** menu.

The integral should appear below the equation you entered.

Press: (ENTER)

Your screen should now show

$$y(t) := e^{-1}\cdot\left(\sin\left(\pi\cdot\frac{t}{8}\right) + \cos\left(\pi\cdot\frac{t}{8}\right)\right)$$

$$y(t) := \left(-8\cdot\frac{\cos\left(\frac{1}{8}\cdot\pi\cdot t\right)}{\pi} + 8\cdot\frac{\sin\left(\frac{1}{8}\cdot\pi\cdot t\right)}{\pi}\right)\cdot \exp(-1)$$

Note that a function can involve a number of variables, and you can perform partial integration with respect to any variable in the function.

Try It Integrate the following functions with respect to *x*:

- $(\sin(x) - \cos(x))^2$
- $\tanh(x)/e^{-x}$

Application 4 COMPOSITE MATERIALS

The one, main goal of any racing team is to win races. To do so, it is mandatory that a race car's engine create as much power as possible. Ultimately, however, it may be better to use the engine's power more efficiently than to generate more power since more powerful engines cost more and tend to be less dependable.

1. Define the Problem

Efficiency involves converting as much of the energy generated by the vehicle's engine as possible into linear kinetic energy of the vehicle. However, a significant amount of the energy generated by an engine is converted into angular kinetic energy of rotating parts. Clearly, the energy that angularly accelerates rotating parts is not available to linearly accelerate the car.

2. Gather Information

One way of minimizing the energy lost to angular kinetic energy is to minimize the moment of inertia of the rotating parts. Here, we will examine the problem of minimizing the moment of inertia of the car's wheels, four of the largest and most massive rotating parts on any racing car.

The moment of inertia of a rigid body is defined as the sum of all its particles' (atoms) masses (m_i) times their distances (r_i) from the axis of rotation:

$$I = \Sigma m_i r_i^2$$

The kinetic energy of a rotating rigid body is defined as

$$K = (1/2)I\omega^2$$

where ω is the angular speed.

This kinetic energy is provided by the engine and is wasted. From equation (1), we see that the lower the moment of inertia, the less the kinetic energy of the wheel. And from equation (2), we see that the smaller the radius and mass of the wheel, the less its moment of inertia.

Clearly, using smaller diameter wheels is a tempting option. They weigh less and will have their mass closer to the center of rotation; both are factors that lower the moment of inertia. However, one can go only so far with this. At some point, the wheels become so small that the suspension that supports them cannot clear the track surface.

3. Generate and Evaluate Potential Solutions

A mechanical engineer who works with a racing team must survey the available tires and wheels. If the engineer recommends a new way of designing and fabricating the wheels, the possible increase in performance must be compared to the cost of doing so.

4. Refine and Implement a Solution

Let's assume the decision is made to go with custom wheels. The size of the wheel will be chosen based on available tire sizes and on the required suspension clearance. The engineer must then choose the material to be used. Often, a stronger, but heavier, material is used for the hub, a less strong, but lighter material for the spokes or area between the hub and outer ring. The lightest, but weakest, material is used for the outer ring. The result is the lowest possible moment of inertia.

Let's assume that the engineer decides to use steel for the hub, aluminum for the spokes, and carbon fiber for the outer ring. The following Mathcad work defines an equation for the moment of inertia for the composite wheel, determines the moment of inertia of the wheel, and lastly calculates the maximum kinetic energy of the wheel, that is, the energy at race speed. The *I* calculation for each component of the wheel (the hub, spokes, and outer rim) reflects the weight (in pounds) and distance from the center of the wheel (in inches).

I for steel hub — I for spokes — I for outer rim

$$= \left[9.85\cdot \text{lb}\cdot(3.2\cdot \text{in})^2\right] + \left[5.43\cdot \text{lb}\cdot\int_{4.4}^{8.25}\frac{r\cdot \text{in}^2}{8.2}\,dr\right] + \left[2.45\cdot \text{lb}\cdot(8.05\cdot \text{in})^2\right]$$

$$= 1.915\cdot \text{lb ft}^2$$ I for entire wheel

$$= \frac{1}{2}\cdot I\cdot\left(293\cdot\frac{\text{rad}}{\text{sec}}\right)^2$$ Definition for angular kinetic energy = (1/2) I w^2

$$= 8.22\ 10^4\ \text{lb ft}^2\ \text{sec}^{-2}$$ Kinetic energy of one wheel at racing speed

This information can be used to determine the effect of changes in the mass and radius of the wheel's parts on the wheel's moment of inertia. It can also be used for comparison with other wheel designs. Clearly, the engineer is searching for a design that minimizes the bottom line, that is, *k*, the angular kinetic energy of the wheel.

5. Verify and Test the Solution

To verify the solution, the engineer might start by testing a prototype wheel on a dynamometer to check the amount of energy it takes to accelerate it. Ultimately, however, the engineer must actually construct a set of wheels, put them on a car, and race. Wheels have been know to break during races. They have also had cracks form in them, allowing the air to escape. Either way, the result is often a wrecked $500,000 car and, perhaps, a fired mechanical engineer.

SUMMARY

This chapter discussed differentiation and integration. We first described how to determine the derivative of a function at a particular point and how to use the Mathcad symbolic processor to determine the function that is the derivative of a given function. Then we explained how to determine the integral of a function over an interval and how to use the Mathcad symbolic processor to determine the function that is the integral of a given function.

Key Words

derivative
differentiation
integral
integration
symbolic processing

Problems

1. Determine the derivative of the following functions at the point $x = 0$:
   a. $y(x) = 4x^2 - 5x + 1$
   b. $y(x) = -3x^4 + 2x$
   c. $y(x) = 5e^{4-3x}$
2. Determine the derivative of the following expressions at the point $x = \pi/8$:
   a. $\sin(2x)^2$
   b. $-\cos(4x)^{-2}$
   c. $5e^{4-3x}$
3. Determine the first, second, and third derivatives of
   a. $y(x) = -3x^3 + x$
   b. $y(x) = -\cos(4x + \pi/2)^{-2}$
   c. $y(x) = 5e^{4-3x}$

4. Determine the first and second derivatives of
   a. $y(x) = \cosh(x^{-2})^2$
   b. $y(x) = \cos(x^{-2})^2 - \sin(x^{-2})^2$
   c. $y(x) = \sin(e^{x} - \pi/4)$
5. Assume the speed of an object is defined by
   $v(t) = -t^2+100t$
   $v(t)$ is in meters per second and t is in seconds.
   a. Determine the time at which the object reaches its highest speed.
   b. Determine the highest speed reached.
   c. Determine the distance the object travels between $t = 0$ and $t = 100$ seconds.
6. Determine the integral of the following functions over the interval $x = -2$ to $x = +2$:
   a. $y(x) = 4x^2-5x+1$
   b. $y(x) = -3x^4+2x$
   c. $y(x) = 5e^{4-3x}$
7. Determine the integral of the following functions over the interval $x = -\pi/8$ to $x = +\pi/8$:
   a. $y(x) = \sin(2x)^2$
   b. $y(x) = -\cos(\pi x)^{-2}$
   c. $y(x) = 5e^{4-3x}$
8. Determine the double integral of the following functions over the interval $x = 0$ to 5 and $y = 0$ to 10:
   a. $z(x,y) = x^2-y^2$
   b. $z(x,y) = e^{-x}+e^{-y}$
   c. $z(x,y) = \sin(x)-\cos(x)$
9. Integrate the following functions with respect to x:
   a. $y(x) = -3x^3+x$
   b. $y(x) = -\cos(4x+\pi/2)^{-2}$
   c. $y(x) = 5e^{4-3x}$
10. Determine the second integral of the following functions with respect to x:
    a. $y(x) = \tan(x)^2$
    b. $y(x) = \cos(x)^2 - \sin(x)^2$
    c. $y(x) = \ln(e^{-x})$

Appendix A

SUMMARY OF MODULE CONTENTS

Array Computations

You can multiply, divide, add, and subtract matrices by using the usual operators. The rules of matrix math must be followed, of course. Mathcad can also compute the matrix transpose, inverse, dot product, cross product, and determinant. A number of statistical functions can be applied to matrices; see Statistics.

Complex Numbers

The roots of many functions are complex. To find these roots, your initial guess must be imaginary. To assign an imaginary value to a variable, follow a constant with the letter i (do not use the multiplication operator). For example:

```
x:2i(ENTER)
root(x^2-x+5,x)=(ENTER)
```

Constants, Predefined

Mathcad knows the values of the constants pi, e, infinity, and %, where e is the natural logarithm base, infinity is stored as 10^{307}, and % is stored as 0.01.

Contour Plots, Creating

To create a contour plot, you must first have a 2D matrix of values to plot. Mathcad considers the indices of the matrix as the distances along the X and Y axes. The elements are the Z-axis values. To create a contour plot of the values contained in matrix **M**:

- Select **Create Contour Plot** under the **Graphics** menu.
- In the placeholder at the bottom of the plot, enter: `M,(ENTER)`

Data Files, Reading and Writing Matrices

The *READPRN*, *WRITEPRN*, *APPENDPRN* commands are used to store, retrieve, and append, respectively, whole matrices. They are used basically in the same way as are the vector commands listed above, only they take a matrix argument rather than a vector argument; however, a range (index) variable is not used.

Data Files, Reading and Writing Vectors

All Mathcad data must be stored in ASCII text data files, with the usual blank, comma, and carriage return delimiters. The *WRITE*, *READ*, and *APPEND* commands are used to store, retrieve, and append data, respectively. To store a vector **X** of, say, 100 elements in a file called DAT, you would type:

```
i:0;99(ENTER)
WRITE(X):DAT(ENTER)
```

To retrieve that data and store it in a vector **Y**, enter

```
Y[i:READ(DAT)(ENTER)
```

All file commands must be typed in all capitals. These commands include *READ*, *WRITE*, *APPEND*, *READPRN*, *WRITEPRN*, and *APPENDPRN*.

Derivatives, Evaluating

To determine a derivative of a function $f(x)$ at a specific x value, assign that value to x and then select the derivative operator (d/dx) from the Calculus Palette on the palette. In the placeholder to the right of the operator, enter the function (either define it here or define it earlier and then enter $f(x)$). In the placeholder in the denominator of the operator, enter x and press (ENTER). For example, to evaluate $f(x) = 3x^2 + 2x + 1$ at x = 2,

```
x:2(ENTER)
f(x):3*x^2(↑)(↑)+2*x+1(ENTER)
```

- Click on the Calculus Palette button on the palette.
- Select the derivative operator (d/dx).
- In the right placeholder, type: `f(x)`

Differentiation, Symbolic

See Symbolic Differentiation.

Documents, Creating

A Mathcad document is the blank page on which the user creates regions (see Regions). Documents can be saved and opened as files. You create a new document by selecting **New** under the **File** menu.

Documents, Editing

To edit any equation, text, or graph region, just click at the point in the region where you want to edit.

Documents, Printing

Select **Print** under the **File** menu.

Documents, Retrieving

Select **Open** under the **File** menu.

Documents, Saving

Select **Save** or **SaveAs** under the **File** menu as appropriate.

Equations, Entering

To enter an equation, simply type it. Remember to use the arrow keys, or the spacebar, to encase parts of the equation in a blue insertion box as appropriate. Use the colon (:) to set a variable of a function equation to a value or expression. Use the equals sign (=) to have Mathcad compute the value of a variable, expression, or function.

Error Messages

Click once on any error message and then press (F1) to get more information on the error.

Functions, Built-In

To select a built-in Mathcad function, click on the Choose Function (showing an *f(x)* icon) button on the tool bar.

Functions, Graphing

Select the graph type you wish under the **Graphics** menu. Double click on the graph or plot to perform formatting functions.

Functions, User-Defined

User-defined functions are created by defining the function in basically the same way as you do on paper. For example: *f(x)*:3*x+1

Graphs, Creating

Select the graph type you wish under the **Graphics** menu. See the specific graph type you want to create.

Graphs, Formatting

Double click on the graph or plot you want to format.

Help

Mathcad 6.0 offers as much help as you could want in getting started on any task, simple or complex. The **Help** menu and the (F1) key call up the usual kind of help found in most Windows programs.

QuickSheets are accessed by either clicking on the tool bar button with the light bulb on it or by selecting **QuickSheets...** under the **Help** menu.

Mathcad's Electronic Books offer graphical examples in many specific areas of engineering. Each Book is a collection of up to several hundred Mathcad worksheets containing text, pictures, formulas, and data. To access the Electronic Books, select **Open Book...** under the **Books** menu, double click on the Deskref.hbk icon, and double click on Table of Contents.

Histograms, Creating

The function *hist*(**int**,**v**) returns a vector that represents the frequencies with which values in the vector **v** fall in the intervals defined by the vector **int**. Both vectors contain real numbers. The returned vector contains integers and one less element than is contained in **v**. For example, to generate 100 random numbers and then divide them into four groups, respectively containing numbers $-50 \le x < -25$, $-25 \le x < 0$, $0 \le x < +25$, $+25 \le x < y+50$

```
i:0;99(ENTER)
V[i:(-50+rnd(100))(ENTER)
j:0;4(ENTER)
INT[j:(-50+25*j)(ENTER)
hist(INT,V)=(ENTER)
```

Histograms, Graphing

To graph a histogram, first create the histogram using the *hist*(**int**,**v**) command. Then create a function of the histogram and, lastly, graph the function as a bar chart. For example, to graph a histogram of the contents of vector **V**, which contains values from 100 to 200, using 10 equal sized groups (100-110, 110-120, ..., 190-200),

```
j:0;9(ENTER)
bins[j:(100+10*j)(ENTER)
hist(bins,V)=(ENTER)
f:hist(bins,V)(ENTER)
```

Now, define the number of bars to be plotted and then create the graph:

```
k:0;10(ENTER)
```

- Select **Create X-Y Plot** under the **Graphics** menu.
- In the X-axis placeholder, enter the index: `k`
- In the Y-axis placeholder, enter the function vector: `f[k(ENTER)`

The graph will display a line chart. To convert it to a bar chart, double click on the graph and in the Formatting dialog box, select *bar* under the Type column. Click on OK.

Integrals, Evaluating

Evaluating an integral of a function *f(x)* involves finding the area under the curve defined by *f(x)*. A beginning and ending value of *x* must be provided. For example, to find the area under the curve defined by $f(x) = 3x^2 + 2x + 1$ over the interval $x = 0$ to $x = 10$,

```
f(x):3*x^2(↑)(↑)+2*x+1(ENTER)
```

- Click on the Calculus Palette button on the palette.
- Select the integral operator that has limits (*a* and *b*).
- In the placeholder to the left of the d, type: `f(x)`
- In the placeholder to the right of the d, type: `x`
- In the placeholder at the bottom of the integral symbol, type: `0`
- In the placeholder at the top of the integral symbol, type: `10=(ENTER)`

The area under the curve should appear to the right of the equals sign.

Integration, Symbolic

See Symbolic Integration.

Least Squares Fits, Graphing

Performing a least squares fit is the same as performing linear regression, that is, finding the straight curve that best fits (or describes) a collection of X-Y data points. Two built-in Mathcad functions are used: *slope*(**vx**,**vy**) returns the slope of a linear function that best fits the data pairs in vectors **vx** and **vy**; *intercept*(**vx**,**vy**) returns the Y-axis intercept of that function. For example, to determine the line that best fits the movement of the monthly average stock price of a company over 6 months, type:

```
i:0;5(ENTER)
month[i:i(ENTER)
price[i:22.5,21.0,18.25,20.75,16.0,18.12
5(ENTER)
f[i:slope(month,price)*month[i +inter-
cept(month,price)(ENTER)
```

Then select **Create X-Y Plot** under the **Graphics** menu. To display both the original data points, connected by line segments, and the straight line of the best fit function:

- In the X-axis placeholder, type: `month[i`
- In the Y-axis placeholder, type: `price[i(↑)(↑)(↑)f[i(ENTER)`

Mathcad Window

At the top of the Mathcad window is the *title bar*, where the word *Mathcad* is followed by the name of the Mathcad document you are editing. Its default name is Untitled:1, but you can give it any legal filename when saving it by selecting **Save As** under the **File** menu. At the left end of the title bar is a button; double clicking on this button (or the X button in the upper right-hand corner if you are using Windows 95) will quit Mathcad.

The second row from the top of the Mathcad window is the menu bar, with **File** at the left end and **Help** at the right end. To pull down a menu, you can either click on it with the mouse or press the (ALT) key followed by the underlined letter in the name of the desired menu. Many of the menu selections can also be activated using the (CTRL) key in combination with another key.

The third row from the top of the Mathcad window is the *palette*. By clicking on the buttons on the palette, you can easily insert arithmetic operators, evaluation and Boolean operators, graphs, vectors and matrices, calculus operators, programming templates, and Greek letters. If you leave the mouse arrow momentarily over any button on the palette, or any other button on the screen, Mathcad will display a concise description of the button's function.

The fourth row from the top of the Mathcad window, called the *tool bar*, contains buttons on which you can click in order to quickly perform some of Mathcad's most commonly used operations.

Below the tool bar is the *font bar*. The font bar contains pull-down menus that allow you to select text styles, fonts, and sizes, and buttons that allow you to select bold, italics, underlining, subscripts, and superscripts.

Matrix, Creating

To create a 2D matrix, select **Matrices** under the **Math** menu. Enter the number of rows and columns you desire in the Matrices dialog box and click on **Create**. Then enter the values in the matrix that appears. Use the tab key to move through the matrix.

Multiple Functions, Graphing

To create a graph that displays multiple functions, the functions must all use the same independent variable. Select **Create X-Y Plot** under the **Graphics** menu. Enter the independent variable in the placeholder under the X-axis. In the placeholder for the dependent variable, to the left of the Y-axis, enter the functions (fully defined or just their names if defined earlier), as you normally would, separating them with commas. For example:

```
widgets:0;99[ENTER]
cost(widgets):1000+10*widgets[ENTER]
income(widgets):12*widgets[ENTER]
```

Then, in the Y-axis placeholder, type

```
cost(widgets),income(widgets)[ENTER]
```

Multiple Vectors, Graphing

To create a graph that displays multiple vectors, the vectors must use the same index. Select **Create X-Y Plot** under the **Graphics** menu. Enter the index as the independent variable in the placeholder under the X-axis. In the dependent variable placeholder to the left of the Y-axis, enter either a definition of the vector contents or the names of the vectors, separated by commas. For example,

```
i:0;9[ENTER]
X[i:81,64,49,36,25,16,9,4,1,0[ENTER]
Y[i:i*2[ENTER]
```

Then, in the Y-axis placeholder, type

```
X[i[↑][↑][↑]Y[i[ENTER]
```

Polar Plots, Creating

To create a polar plot, you usually begin by defining a range variable, often ranging from 0 to 2*pi radians or 0 to 360 degrees. Then you define the function either directly or by filling a vector with appropriate values. Lastly, you select **Create Polar Plot** under the **Graphics** menu and enter the independent and dependent variables. For example, to create a polar plot of the function $f(\text{angle}) = -1.1^{0.25\theta}$, angle = 0..360 degrees,

```
i:0;360[ENTER]
```

To convert the degrees defined by *i* to radians, type:

```
angle[i:i*2*[CTRL]p/360[ENTER]
f[i:-1.1^angle[i[ENTER]
```

- Select **Create Polar Plot** under the **Graphics** menu.
- In the placeholder at the bottom of the plot, enter: `angle[i`
- In the placeholder at the left of the plot, enter: `f[i[ENTER]`

Random Numbers

Use the *rnd* command to generate random numbers; *rnd*(*x*) returns a random real number between 0 and *x*. To generate a random integer, use the *floor* command with the *rnd* command. For example, to create a 10-element vector **X** filled with random integers from 100 to 19, type

```
i:0;9[ENTER]
X[i:100+floor(rnd(100))[ENTER]
```

Regions, Creating

There are three types of regions in a Mathcad document: *equation, text,* and *graph.* The default region type is equation. If you simply begin typing text when the cross-hair is present, Mathcad will assume you are entering not text, but a variable name as part of an equation. To create a text region, you must first type a double quote ("). To create graphs and plots, you select a graph type under the **Graphics** menu.

Regions, Moving

To move one or more regions, drag the mouse across them until a dashed box appears around each region. Then click in any of the regions, holding the mouse button down, and drag the region or regions to any new location. Pressing the [BKSP] key will delete any selected regions. Pressing [CTRL]+s will separate any regions that overlap.

Running Mathcad

To run Mathcad, you must first run Windows or Windows 95. If a Mathcad Program Group icon is present, you must double click the mouse on it to gain access to the Mathcad icon. Once you see a Mathcad icon, double click on it to run Mathcad.

Scatter Plots, Creating

A scatter plot differs from surface and contour plots in that it can plot multiple *z* values at a given *x*,*y* position. To do this, three vectors are used instead of a single 2D matrix. The vectors, respectively, contain the *x*, *y*, *z* positions of each data point. Of course, the three vectors must be of equal length, and a given *x*,*y* position can have more than one corresponding *z* position. Assuming you have three vectors, **X**, **Y**, **Z**, that contain the *x*,*y*,*z* positions of the data points to be plotted:

- Select **Create 3D Scatter Plot** under the **Graphics** menu.
- In the placeholder at the bottom of the plot, type: `X,Y,Z`(ENTER)

Double clicking on the plot allows you to manipulate many formatting options.

Solving Single Equations

To determine the value of any function *f(x)* for a given *x*, for example, 3, type:

```
f(3)=(ENTER)
```

To find the root of any function *f(x)*, first give Mathcad an initial guess for the root's value and then use the root command. For example:

```
f(x):5*x-4(ENTER)
x:0(ENTER)
root(f(x),x)=(ENTER)
```

Solving Sets of Equations

To solve a set of equations, use a solve block, which is started by entering the key word `given`. Start by providing initial guesses. For example, to find the intersection of two parabolas, type:

```
x:1(ENTER)
y:1(ENTER)
given(ENTER)
y(CTRL)=(x-5)^2(ENTER)
y(CTRL)=3*x^2+4(ENTER)
find(x,y)=(ENTER)
```

Remember to hold the (CTRL) key down while pressing the equals sign (=) after each *y*.

Sorting

The function *sort*(**v**) returns the vector **v** sorted in ascending order; *reverse*(*sort*(**v**)) returns **v** sorted in descending order.

Statistics

The most commonly used statistical functions are *mean*, *stdev*, and *var*, all of which operate on vectors. For example, to find the mean, standard deviation, and variance of the real numbers contained in vector **V**, type

```
mean(V)=(ENTER)
stdev(V)=(ENTER)
var(V)=(ENTER)
```

Summations and Products

- To insert a summation operator, type: (CTRL)+ (SHFT)+4
- To insert a summation operator, type: (CTRL)+ (SHFT)+3
- Fill in the placeholders as needed and, to evaluate, type: =(ENTER)

Surface Plots, Creating

To create a surface plot, you must first have a 2D matrix of values to plot. Mathcad considers the indices of the matrix as the distances along the X and Y axes. The elements are the Z-axis values. To create a surface plot of the values contained in matrix **M**:

Select **Create Surface Plot** under the **Graphics** menu.

In the placeholder at the bottom of the plot, enter: `M`(ENTER)

Symbolic Differentiation

To perform the symbolic differentiation of a function *f(x)*, click on any instance of *x* in the definition and select **Differentiate on Variable** under the **Symbolic** menu. For example:

```
f(x):3*x^2(↑)(↑)+2*x+1(ENTER)
```

- Click on either of the two instances of *x* on the right side of the equation.
- Select **Differentiate on Variable** under the **Symbolic** menu.

The derivative should appear below the function.

Symbolic Integration

To perform the symbolic integration of a function *f(x)*, click on any instance of *x* in the definition and select **Integrate on Variable** under the **Symbolic** menu. For example,

```
f(x):3*x^2(↑)(↑)+2*x+1(ENTER)
```

- Click on either of the two instances of *x* on the right side of the equation.
- Select **Integrate on Variable** under the **Symbolic** menu.

The integral should appear below the function.

Tables, Creating

To create a table, start by creating a range variable, for example, by typing

```
i:1.0,1.1;10(ENTER)
```

which defines *i* as ranging from 1.0 to 10.0 in 0.1 steps. (Remember that the second value in the series (the one that follows the comma) defines the second value of the range variable, not its step size.) Then define a function in which the range variable is the independent variable, for example, by typing

```
f(i):i^2(ENTER)
```

To see the table, type: *f*(*i*)=(ENTER)

Text, Entering

Type the double quote character (") and start typing. Click outside the insertion box to quit entering text.

Units, Using

To add units to a constant or variable, multiply the constant or variable by the abbreviation for the units. For example, to define the acceleration due to gravity, enter:

```
acc:32*ft/sec^2(ENTER)
```

To get a list of available units, click on the Units button (showing a measuring cup icon) on the tool bar. Mathcad tracks all use of units and gives an error message if they are not consistent.

Variables

Rules that guide the selection of variable names are the same in Mathcad as in most other programming languages. Use the colon to assign a value to a variable. If Mathcad needs you to provide the value of a variable, it will highlight variable name. Be sure you define all variables and functions above or to the left of where you use them.

Vectors, Creating

To create a vector, start by creating the index as a range variable (see Creating Tables). By default, the index should start at zero, and it must increment in whole numbers. You can create the vector by manually entering its elements, separated by commas; for example, by typing

```
i:0;7(ENTER)
X[i:0,1,4,9,16,25,36,49(ENTER)
```

You can also define a function that is based on the index; for example

```
X[i:i^2(ENTER)
```

Lastly, you can use the **Matrices** dialog box by selecting **Matrices** under the **Math** menu. In the dialog box, type the number of rows (elements) you want, enter the number of columns as 1, and then click on the Create button.

Vectors, Graphing

To graph a vector, select **Create X-Y Plot** under the **Graphics** menu. Then enter the range (index) variable as the independent value (under the X-axis) and the name of the vector, indexed by its index variable, as the dependent variable (to the left of the Y-axis). For example, if *i* is the index and **X** is the vector, the dependent variable should be entered as X[i.

Vectors, Reading and Writing Data Files

See Data Files.

Index